Basic

Lathework

Basic
Lathework

Stan Bray

SPECIAL INTEREST MODEL BOOKS

Special Interest Model Books Ltd
P.O.Box 327
Poole, Dorset
BH15 2RG
England

First published 2010
© 2010 Stan Bray

ISBN 978 185486 261 7

www.specialinterestmodelbooks.co.uk

Printed and bound in Great Britain by Martins the Printers Ltd

Contents

Introduction

There has always been an interest in using a centre lathe and there have always been many home workshops, most were garden sheds. Things have and are changing, the garden shed is now much more sophisticated and many workshops are no longer in the garden at all. As with all things the years have brought improvements to equipment and the days of the treadle operated lathe or the one powered by a flat belt from an overhead shaft. For many years lathes have been self-contained machines driven by an electric motor. Progress in their design is still taking place and our forefathers would probably be at a loss to know how to work the modern machines.

Another change has also taken place and that is in the knowledge of many who have or want to have a lathe. At one time Britain was a manufacturing country, the majority of people worked in factories and to them the lathe would have been a familiar machine even if they did not actually operate one. Until the late twentieth century metalworking was part of the school curriculum and youngster gained hands on experience of operating a lathe. In a very short period of time it all changed, metal working was no longer a subject taught at most schools, factories closed down and the average persons' knowledge of engineering of any sort became nil.

Strangely in spite of this the urge to own and operate a lathe appears not to have diminished but instead to have increased, hence the healthy market for machines.

There is little doubt that many, perhaps even the majority of lathes that are installed at home are never used to their full potential, because their owners do not have sufficient knowledge to do so. It is hoped that this nook will be of assistance to such people and widen their knowledge, thus allowing even greater pleasure to be obtained for ownership of a machine. Of course there are other books on the techniques of using a lathe, some written many years ago. Things have changed with machines becoming more sophisticated and hopefully this is reflected in these pages. At the same time more traditional methods have not been forgotten because in spite of such things as digital read outs and computer control gradually taking over the basic principles remain.

Working on the principle that a picture is worth a thousand words, as many illustrations as possible have been included. It is far easier to look at how a thing is done than to read about it and try and visualise it, It is hoped therefore that the book will give pleasure as well as imparting knowledge to those who read it.

Stan Bray 2010

Safety

There are dangers associated with operating a lathe but then there are dangers with everything we do in life and so safety is largely a matter of using the same common sense to operating a lathe that we use at other times. Because metal is being cut there is a danger is it flying up and possible getting in ones eyes, so it makes sense to wear safety glasses. These can be purchased in various forms and if glasses are worn it is possible to obtain safety glasses to ones own prescription. There is a danger of catching oneself on the revolving work if not being careful so here again it is a good idea to use a chuck guard on the machine. None of the pictures in the book show chuck a guard in place of course, if they did it would not be possible to see what was going on so the guard was deliberately moved for the photograph. At one time long hair was a problem as it was likely to catch in the revolving belt of the lathe. Unless one is operating a very old type of machine without a belt or chuck guard the problem is unlikely to be present but it might be wise for anyone with long hair to do something about it before using the lathe. It is not a good idea to operate the lathe wearing ones carpet slippers. Apart from the fact that there is a possibility of them being splashed with oil and it subsequently getting on the best carpet, there is the danger of dropping a piece of metal and inevitably it will land on ones foot. Oil does splash about and although a chuck guard should prevent most of it from getting away it is not ideal to wear ones best suit, far better to have some form of overall. We could of course go on forever about safety measures, just use common sense and dress accordingly, nobody wants to spoil the pleasure of using the lathe by having to have a major change of dress every time.

Chapter 1

The lathe explained

No doubt everyone taking the trouble to read this book will know that a lathe is a machine, with a rotating spindle that is used for shaping, wood or metal. We are concerned here with metal turning machines, (turning being the term applied to most lathe operations,) rather than wood turning. The lathes used for that purpose are of a much simpler form of construction than those designed for metal work.

Let us start by familiarising ourselves with the basic parts of the machine and although there have been many developments over the years the overall layout of the lathe varies little from what it did a hundred or more years ago.

The bed
The main structure of the lathe is the bed, a longitudinal piece of metal, usually cast iron, on which everything else rests and works. The bed will have some form of support, to raise it off the bench or stand on which it is to rest. These usually take the form of a couple of squat pillars of a shape that is sympathetic to the shape of the bed itself. The top face of the bed is smooth as it is used as a slide way, it may be flat, or it may have either a single or double vee shape cast along, it. The vees will be ground smooth as they are required for the movement of other parts of the machine. Most lathe beds have an open section along the centre, to facilitate the fitting of other parts, there are some examples however where the bed is box shaped instead.

The operators' side of the bed will have several attachments, these include a rack that is used to move the saddle and assuming the machine is equipped to do screw cutting, a long thread, known as the lead screw. The lead screw also provides a means of allowing the saddle to move under power.

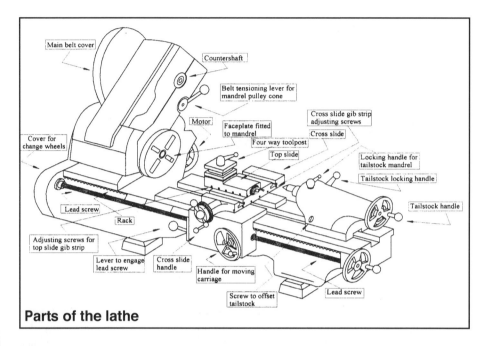

Parts of the lathe

Labels in figure:
- Main belt cover
- Countershaft
- Belt tensioning lever for mandrel pulley cone
- Cross slide gib strip adjusting screws
- Motor
- Faceplate fitted to mandrel
- Cross slide
- Cover for change wheels
- Four way toolpost
- Top slide
- Locking handle for tailstock mandrel
- Tailstock locking handle
- Lead screw
- Tailstock handle
- Rack
- Adjusting screws for top slide gib strip
- Lever to engage lead screw
- Cross slide handle
- Handle for moving carriage
- Lead screw
- Screw to offset tailstock

The headstock

The parts that rotate are mounted in or on the headstock, which is a hefty casting standing on top of and attached to the bed. The size of this will depend on the design of the machine and how sophisticated it is. The headstock will hold the speed changing mechanism and most importantly it will hold the mandrel, or spindle, as it is now more commonly known. The probable reason that it does tend to be referred to as the spindle is because we use mandrels in some instances for machining purposes and describing both the one being worked on and the lathe spindle by the same term can be extremely confusing. The spindle is the part that rotates and how this is achieved is explained elsewhere, the headstock is cast in such a way that the

bearings for the spindle can be fitted and these can take several forms. Some lathes have plain bronze bearings that are split across the circumference. To mount the spindle a removable section of the headstock can be unbolted and the spindle laid into the bottom half of the bearing that fits in a suitably machined recess in the casting, the top half of the bearing is put on and the upper half of the housing bolted into place. Sometimes shims are used between the two halves of the bearing, if wear occurs these can be removed and so allow the bearing to remain perfectly serviceable. Removing the spindle is of course the exact reverse of fitting it.

Some modern machines have the mandrel fitted with taper roller bearings; this gives a nice smooth running and long lasting fit. There are also examples where one end of the mandrel has a bronze bearing and taper rollers are used at the other end. Every designer appears to have his or her own ideas on which is the best system and it is just a matter of personal choice.

Some very old machines did not have any form of removable bearing, the spindle simply running in the cast iron of the headstock. It is most unlikely one will come across this arrangement unless deliberately searching for an old machine with the intention of restoring it.

The mandrel

As already explained this is now frequently referred to as the spindle; at one end it will have facilities for holding chucks, faceplates etc. The other end will almost certainly be fitted with a gear wheel that is part of a series, ultimately connected to the lead screw and used for obtaining longditudinal motion for the saddle. The spindle is usually hollow and at the end that holds the chuck there is a taper, the size and type of this will depend entirely on the type and make of lathe. It is worth mentioning that the larger the bore of the spindle the better, as it allows long lengths of metal to pass right through for machining and if it is fairly large so much the better. The spindle will also hold the system used to rotate it and in most cases provision to alter the rotational speed. Usually this will be in the form of a set of pulleys in the form of a cone, There are just a few and it is very few, examples of older small lathes that have solid spindles rather than hollow ones, it would be wise to ignore any such machine as the solid spindle imposes considerable restriction on the work that can be done. On most lathes there will be

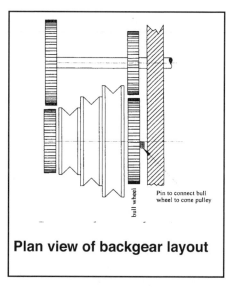

Pin to connect bull wheel to cone pulley

bull wheel

Plan view of backgear layout

facilities to oil the spindle bearings, this will only be absent in the case of a machine with sealed bearings.

Spindle speeds

Most older machines will use a system whereby a belt is moved from one pulley to another in order to change spindle speeds. Very old machines have a system that uses a flat belt, newer ones use a vee shaped one. The system on both types is virtually the same, there is a countershaft that is belt driven from a pulley on the motor and the shaft contains three or four pulleys of different diameters, arranged in ascending order of size known as a cone pulley. The lathe spindle will also have a cone pulley arranged in the opposite order to those on the countershaft. A belt connects a pulley on the counter shaft to the opposite one on the spindle and by changing the belt from one pair of pulleys to another the spindle speed will be changed. It is an idea that has stood the test of time and is still to be found on many new machines that are sold today.

There is absolutely no reason whatever why a belt from the motor cannot drive the shaft directly, except of course that it severely limits the rotational speeds available to the operator. Even three speeds is limiting and so we will mostly find that in addition to the change pulleys there will be a system called the back gear. This consists of a shaft that is driven by reduction gears from the spindle shaft. In a normal position it is disconnected and has no effect on the spindle speed. To select it, the gear which is normally connected to the cone pulley is disengaged. This is usually achieved by moving a pin, and a shaft that has a gear wheel at either end is moved to mesh with the gear wheel at the end of the shaft where the smallest pulley is usually situated. The movement of the shaft also engages a gear at the opposite end with the bull wheel, which is a large diameter gear wheel, thus giving the required reduction. The cone pulley is of course still connected to the shaft and therefore can be used in the normal manner allowing a variety of the slower speeds to be available.

Some machines have the motor and countershaft housed in the lathe cabinet, it makes no difference to the operation of the machine, but looks neater as well as saving space. As an additional way of changing speeds some lathes are fitted with a double pulley on the end of the countershaft that matches a double pulley on the motor shaft. This too means that changing the belt between motor and countershaft can double the number of speeds obtainable by changing the belt.

While it is effective, changing belts can be time consuming and a nuisance and a good proportion of lathes now have what is known as a geared head. This is literally a gearbox that replaces the countershaft and is driven by the motor. Changing spindle speeds becomes a simple matter of moving a couple of controls on the headstock.

The latest ideas on varying spindle speeds use variable speed motors, driving via a belt directly to the spindle. The idea finds favour particularly with the small or mini lathes as it is particularly suited to the small motors involved. It can also be found on some quite large models and these usually have a built in tachometer that shows the speed of rotation of the mandrel,

The saddle or carriage

Referred to as a saddle because it straddles the lathe bed just as a saddle straddles the back of a horse and also alternatively as the carriage because it moves along in a similar fashion to a carriage, it could be said to control most machining operations. Fitted to it is a vertical plate known as the apron and on that we find various controls for its operation. These include a hand wheel for traversing the saddle along the length of the bed in either direction. This is done via a gear wheel that is engaged with a rack, screwed to the side of the bed, usually just under the top surface. Attached to the bed below the rack will be the lead screw, this rotates when connected to the spindle and according to the ratio of the gearing between spindle and the lead screw, the saddle moves a certain distance for each rotation of the mandrel. It is usually possible to reverse the rotation of the lead screw so that the saddle can move in either direction, the control for reversing the movement is generally situated near the base of the headstock. On older lathes it will most likely be the case that there is no method on the apron of mechanically measuring the distance the saddle has travelled, modern machines are likely to have a digital arrangement for doing so. The lead screw is the whole length of the lathe bed, it has to be so in order to carry out its purpose of transporting the saddle and on some lathes there will be a handle on the end of the screw, it will be graduated and therefore offer some means of knowing how far the saddle has moved. The connection of the saddle to the lead screw is obtained via a lever and this closes a half nut over the screw. A half nut is literally what it says, a nut that has a thread to match that of the lead screw and has been cut in half. The action of the lever is to close the halves together, thus taking a grip on the lead screw.

Later in this book there will be a description of methods of obtaining a much more accurate measurement than can be obtained with the handle. Really old lathes, without facilities for screw cutting or automatic traverse of the carriage will not have an apron, just a bracket to hold the handle that moves the carriage along.

The cross slide

Mounted on top of the saddle is the cross slide, so called because that is literally what it does, slides across the line of the bed. Movement is obtained by winding a handle backwards or forwards that operates a screw, also called a lead screw, which fits a nut attached to the slide, thus creating the necessary movement. The screw is of either a square or acme formation in order to supply the required power and will be of steel. The nut ideally will be of bronze, although there are some lathes that have lead screws made of white metal. The slide casting usually has a collar with a mark that lines up with graduations on a collar attached to the lead screw allowing one to know exactly the distance the cross slide has moved. In many instances the dial is adjustable to allow it to be reset to any required figure without disturbing the position of the cross slide. More modern lathes are likely to have this arrangement replaced by a digital read out arrangement, giving greater accuracy.

Although lead screws are the most common method of driving mechanisms of this sort they do suffer from a failure called backlash. To explain this it should be pointed out that a perfect thread is not a practical proposition. If the two parts were made a perfect fit they would be unable to move and so there must be some slackness. This slackness means that there is a small movement between the two when the handle is rotated, so the first movement of the handle does not actually move the slide at all, it is only when the two parts of the screw mate with each other that movement can take place, therefore we have a dead spot. Although the reading on the dial may show that the slide has moved a tiny fraction it will not have done so, how great this error is will depend on how good the fit of the two parts of the screw are. The usual way of trying to compensate for this is to wind the handle in the reverse direction until it can be felt to be free and then wind it to a point where it is in contact with the other part of the thread. The reading of the dial is taken at that point, as it is from there that the screw is actually being driven.

Many modern machines do not rely on a lead screw dial reading for measurement, even though they are operated with the nut

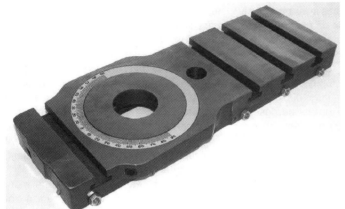

A dismantled top slide showing angle graduations for taper turning and tee slots for work and equipment holding.

and screw. They are fitted with a digital readout, the figures on which only move when the slide itself moves; therefore the question of backlash does not come into it.

Most cross slides will have a number of tee slots across them that allow other items to be bolted down if required. The slide is in two parts in order to allow the necessary movement. The two sections slide together in dovetails, which are angled sections and in order to ensure that there is no unwanted sideways movement or shake one of the dovetails will be fitted with a gib strip. This is a piece of mild steel of suitable size and it is pushed against the sliding surface by means of screws, situated along the edge of the cross slide. These in turn are fitted with locking nuts in order to prevent them from coming loose during operations. To adjust them the nut has to be undone and the screw tightened until the slide can move but there is no side play. If adjustment is required all the screws should be dealt with, rather that trying to adjust a single one. The lock nuts are loosened and the screws tightened to a

point where the gib strip is just making contact. The screw should then be held firmly in position with a screwdriver to prevent it rotating, while the lock nut is tightened. The slide should be wound in and out during adjustment so that it can be ascertained that there are no tight spots. It is an operation that at first can be a little difficult but in time becomes second nature.

The length of the cross slide often is something of a problem and there are times when one might often wish that it were longer. Having a longer one allows tools to be mounted on the opposite side of the mandrel which, particularly for parting off operations as well as any milling that need to be done can be extremely useful. Some lathes do come with a longer cross slide, in other cases it may be possible to buy a suitable one. Making one is a common project for lathe owners.

The top slide

Situated on top of the cross slide, it is usual to mount the tool post on the top slide. It is a small version of the cross slide, except that it is possible to rotate it and use it for cutting work at an angle. To facilitate this there are graduations on the base. Like the cross slide it is usually made of cast iron and of dovetailed construction. While it is useful it also has its disadvantages as the additional height as well as the fact that it is yet another bit added to the saddle means that it could create a certain amount of instability, resulting in uneven finish on work. Some people use it for normal parallel turning operations along a length of material and it is most certainly not suited for this purpose. To use it in this way means relying on the accuracy of the graduations and the slightest misalignment will, result in a tapered finish instead of a parallel one. It too works on the principal of using gib strips for adjustment and has graduations in the same way as the cross slide.

The tool post

Usually situated on top of the top slide will be a means of holding the tools required for machining operations. It may be nothing more that a simple clamp arrangement or something much more sophisticated. It is a very important part of the machine and deserves a lot of discussion and therefore will form a chapter in its own right. The above paragraph relates how the top slide is less stable than the cross slide and so there is something to be said for mounting a tool post directly on to the cross slide rather than using the more normal method of securing it.

The tailstock

As implied by the name, the tailstock fits at the opposite end of the lathe to the headstock; it slides on top of the bed and can be locked in position with a lever that operates a cam. It is a casting with a hollow spindle that is non-rotating and the spindle will have a means of holding certain types of tools that usually incorporate a tapered end section or morse taper. Morse tapers are produced in a range of sizes indicated by numbers. The size required will be matched to the size of the lathe, so for example a mini lathe may well have a No. 1 taper, a medium lathe a No. 2 and so on. The maximum that is likely to be found in a home workshop is a No. 4 and even that would be quite rare. Two or three are the most usual sizes. Some very small lathes have a half taper, which is only really a No. 1 reduced in length, for convenience. A means of moving the mandrel will be fitted and as a rule it will be a handle, the movement being transferred via a screw. In just a few cases the movement will be controlled by a lever arrangement. In addition to being able to lock the tailstock in position there is a means of locking the spindle or mandrel and again this is usually a small lever. Measurements of the amount the spindle has travelled in most case will by means of a series of graduations on the spindle itself, but there are examples where a graduated collar is fitted to the hand wheel.

That covers the basic parts of a lathe; more details will emerge during the course of the chapters that follow. There are other attachments that are desirable but not essential and different makes of lathe will have differing layouts. However just knowing the various parts and attachments is only half the battle, it is also necessary to have some idea of problems that might arise.

Probably the most common is vibration, something that many people do not suspect will be a problem at all. Ninety nine percent or more of lathes are sturdy

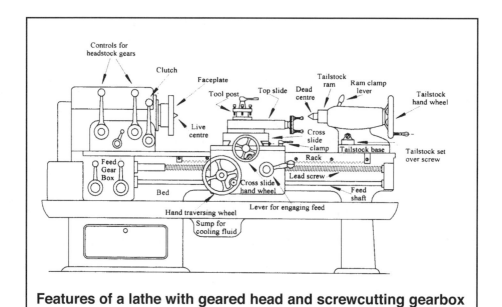

Features of a lathe with geared head and screwcutting gearbox

well built machines. The stoutness of construction will vary according to the size but taking that into account, it is hard to believe that there could be any flexibility in the machine itself, assuming that it has been properly secured to a stout bench or stand. This is not so and it is necessary when using the machine to allow for any problems that might arise. What then is it that will create the movement? Well if we stop to think about it there must be some movement in the main bearings, if this was not so and they were absolutely tight they could not rotate. It may the tiniest amount but it is there. The same applies to the saddle, cross slide and top slide, they too would not move unless there was a tiny amount of play. Already things are adding up say a thousandth of an inch in each bearing and the same in the other

components and already we have a very minimum of five thousandths of an inch which may not sound a great deal. Supposing a length of bar about 6ins (100mm) is put in the chuck (and incidentally there will be some more minute play there as well), at the extreme end of the bar the play will be considerably magnified. This will not mean that here is any problem, with the machine, it is just a fact of life. What the operator has to do is to be aware of it and tailor his or her working practices accordingly. It is no good taking a deep cut from outer end of the bar, the result will be a very rough finish so it means that cuts taken well away from the chuck need to be much smaller than those taken near it,

How about the machine itself, will the bed and headstock have flexibility? To find the answer to this mount a piece of square

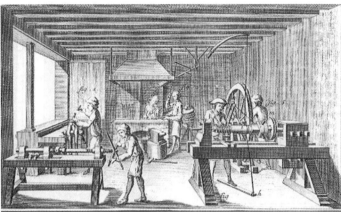

An engraving showing what is probably the earlies type lathe. Known as a pole lathe, such machines were easy to construct and were in use for many years.

bar in either the four jaw chuck or on the faceplate, start the machine and lightly run a cutting tool along it, this can be done quite close to the chuck. If while the tool is contact with the bar a hand is placed on the bed some distance away from the revolving work, so one doesn't injure oneself, it will be found that the bed vibrates quite considerably when the tool makes contact with the edge of the bar. Once again there is nothing untoward in this but it is something we must be aware of and make allowances for, it is obvious that good results cannot be obtained otherwise.

The better the quality of the machine and the heavier its construction the less vibration there will be, but rest assured there will still be a certain amount and machine operations must be tailored with this in mind. It is also absolutely essential to ensure that the machine is securely fastened to a stand or bench, which should in turn be fixed to a wall or floor, anything in fact to avoid vibration.

Chapter 2

Choosing a machine

This chapter is designed to help those that are about to buy a lathe, whether it is new or used. The prospective purchaser is advised to read further into this book before going ahead because while here are detailed the things to look for and look out for, all of the equipment referred to is dealt with in depth in the chapters to come.

There are three main points to be considered when buying a lathe. What will it be used for? How much does it cost and how much space is there in the workshop for it?

For someone with limited space a variety of small lathes are available. This photograph shows a Proxxon which is an ideal example. It is quite capable of doing very good work and at the same time is small enough to be put away in a cupboard or even a deep drawer when not in use. It has an electronic speed control system and it is possible to purchase a whole range of accessories in order to increase the versatility of the machine. (Photograph courtesy of Proxxon Tools)

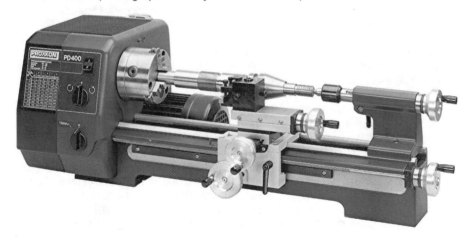

Another small lathe that is useful where space is limited is the Sherline. A number of dedicated users of this make of lathe have organised a web site where help and assistance can be obtained. This lathe also has electronic speed control and various attachments are available as extras if required.

Size is an important consideration, and the machine must be of a suitable size to allow it to be used with ease when it is installed. It is always a good thing to have a lathe that will be capable of being used for work somewhat larger than originally envisaged. Room must be left to operate it and it will be necessary to get at either end of the machine and it is therefore of little use to buy a model that will occupy the whole length of the workshop space. Apart from the difficulty of using it in those conditions, it may be that at a later date one wishes to install an additional machine. This will not be practical if a machine that is too large is purchased. It must therefore be a case of as large as possible without

going over the top. While for most people the advice to buy as large a lathe as possible make sense, someone who is just dedicated to making clocks, or perhaps fittings for model boats, will have no need for a large lathe. He should therefore select one of the smaller models, making sure it has as high a specification as possible.

Lathes are usually sold as being of a certain centre height, and distance between centres. If we take the distance from the bed to the centre of the mandrel, that is the centre height, sometimes the term swing over the bed will be used instead of centre height. It is a term more likely to be found in the USA than in Britain, but in all probability will apply if purchasing an

The Myford range of lathes has been popular amongst model engineers for many years and are particularly popular in Britain where they are manufactured. The photograph shows a Super 7, which is near the top of the range of models. A particularly wide range of accessories are available to suit all the models and form many years modelling magazines have offered constructional descriptions of home made extras.

American lathe. In that case the figure quoted will be double the centre height, it will to a large extent depend on how the supplier likes to describe the machine. The distance between centres more or less speaks for itself, being the distance measured between a centre in the mandrel and one placed in the tailstock when it is at the extreme end of the bed. If that distance is to be important for the work the machine is to do, it is as well to keep in mind that when a chuck is in use the distance will be reduced, because of the overhang of the chuck

There is quite a range of small machines available for those that have definitely decided that such a machine will be suitable. These vary from a pure watchmaker's lathe to one of the small general-purpose machines. Most of these have a centre height of around 75mm or 1-1/2 in, the distance between centres being about 250mm or 6 in It does vary considerably according to the make of the machine. All sorts of accessories are likely to be available, for this type of machine; this is less likely with a pure watchmaker machine.

If the intention is to make large-scale models, or do major restoration on full-sized traction engines or large motor vehicles then obviously a large lathe is desirable. Generally speaking, this can mean

A slightly larger lathe, the Emco Compact 8 one of a range of models available from a popular European manufacturer. A very high precision machine for which spares and attachments are easily available. (Photograph courtesy of Pro Machine Tools Ltd)

anything upward from a centre height of 5in or 125mm and a distance between centres of 24in or 600mm in spite of their size they are also capable of doing small work as well, even though doing this might be more difficult.

It is probably fair to say that the majority of machines that are sold fall somewhere between these two categories, particularly as far as model making is concerned. They have a centre height of about 3-1/2in or 80mm and a distance between centres of around 18in or 500mm. As with other models this figure will vary according to the manufacturer. In some instances, the lathe will have a gap just in front of the mandrel, that allows larger material to be machined, the amount of increase depending on the depth of the bed. Some large machines have a removable bed section that allows a gap to be made; the removable section

may remain in place for normal work. This size of machine will cope very well with a large range of jobs, including some restoration of motorcycles and motorcars as well as the construction of decent sized models.

Together with the question of cost, which must be a personal matter anyway comes that of whether to purchase a new or used machine. It is nice to start with a nice brand new and shiny machine, although the shine rapidly disappears if a lot of work is carried out. There is though no reason why if the selection is made with care an equally good used machine cannot be obtained. There is always a wide range of used machines available either being sold by the owner, or through a dealer. The main difference between buying a new machine and a used one is whether or not the purchaser knows the difference

A lathe of far eastern manufacture, in this case marketed by Chester Machine Tools It is one of a very wide range of models that are available, sizes ranging from very small to very large. There are numerous specifications, from basic to Hi Tech. Similar if not quite identical models are available from a number of these importers, most are finished in the colours peculiar to the company that has imported them.

between a good or bad machine. If a new one is selected, it is reasonable to expect that the machine will be a good serviceable one and the dealer will be legally obliged to ensure that it is in good working order.

Even if buying new it is still necessary to be careful, because of the ease of obtaining imported machines. Companies that have no real knowledge of their product, such as DIY shops nowadays sometimes retail lathes. The people selling them do not know how to use them or anything about them and purchasing a machine from this sort of supplier is not to be recommended, particularly if the purchaser also has little or no idea of what he or she might require. Certain equipment will be necessary in order to use the machine and one bought under the above

circumstances may not have what the purchaser really has need of. The equipment to expect and what will be required will be detailed shortly.

The suppliers in the Far East are only too happy to allow their machines to be sold by a company under it's own brand name. These machines are often painted in the colours required by the retailer and have the retailer's name on them. The result of this is that there are identical machines available, sold under different names and the prices vary considerably. It is therefore wise to get us much information as possible from as many dealers as possible in order to make a comparison. Reputable dealers will check all machines before they are sold, some even offer a service of making extra checks for a small addition to the price. The

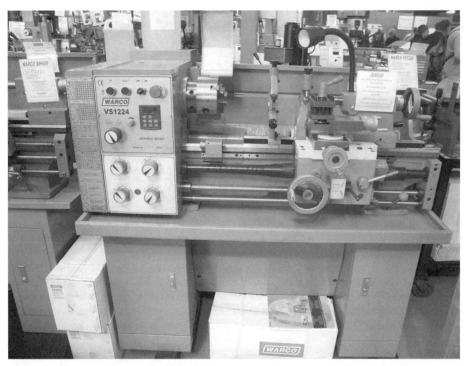

Another example of a lathe imported from the Far East in this case the importer is Warren Machine Tools and the machine is in their colours. All the higher ranges of these imported machines have a good specification.

type of dealer referred to above that has no in depth knowledge of the machines is less likely to have the know how or ability to check the machine over. The result will be a machine that may or may not have had faults before it was despatched and anyway has travelled thousands of miles, possibly under adverse conditions. This is now sold to an unwary client as an accurate machine in good working order. This is not always the case. This does not mean that all machines manufactured in the Far East are rubbish, there are some very good examples, it is simply a matter of selecting

with care and not buying a lathe just because it is painted in nice bright colours.

It must be emphasised that while it is necessary to beware when buying a machine from the Far East, providing it has been thoroughly checked over the machines are very reasonably priced and many have quite a high specification. Of course new machines that are on sale do not all come from the Far East, they are imported from all over the world, some are made in Britain while others are made on the European mainland and while generally the prices of these machines are somewhat higher, they

A popular machine in industry and yet not too large for many home workshops is the Colchester. The one shown is a modern type, older models frequently come on the market as used machines.

usually represent excellent value.

Necessary accessories

Certain accessories will be required in order to operate a lathe successfully and new machines should be supplied with them. It may not be so if a used machine is purchased as the person selling it might decide to keep certain items in order to use them and this should be taken into account when buying the machine. To be useable for a reasonable variety of work all machines should come with a motor and switchgear and should have two centres (these are described elsewhere) and a faceplate. It is possible although not desirable to work with that bare minimum of equipment but most people will want rather more and the first priority will be a chuck. Most new lathes are sold with a three -jaw self-centring chuck and while these are convenient to use they are

very limited in what they can actually be used for. Basically all that such a chuck will do is to hold round or hexagonal bar, which means all that one can actually do is to machine the bar to a smaller diameter. Although to the newcomer they are initially difficult to use a four jaw independent chuck is a great deal more use. All types of chuck will be dealt with at length later on. The other item that is almost indispensable is a tailstock chuck, which is like the chuck on a drilling machine, but designed to fit in the tailstock of the lathe. Last but far from least check that the machine has a suitable tool holder.

Screw cutting facilities are a desirable feature, although it must be said that some purchasers will use the machine for work that will never require such a facility to be used. An automatic feed arrangement for the saddle and capability to reverse the

The Harrison is a large lathe but is popular with model makers building large models such as large-scale locomotive and traction engines. Used models are often sold at comparatively low prices. A lathe of this size can create difficulties for some users as it requires either considerable strength or some form of lifting equipment when it comes to operations such as changing chucks.

direction of both mandrel rotation and saddle feed are also something that should be high on the list of priorities. It is possible to work without them but it seems almost inevitable that the time comes when they would be useful. On many lathes these operations are carried out by using a set of change wheels or gears and if this is so it should be established that the full set of these are included. Trying to purchase them later could prove to be very difficult as well as very expensive.

Buying a used machine

If you have no knowledge of lathes, try and take someone who has some knowledge with you to view the machine it is proposed to purchase. Do not be in a hurry to buy, make sure that the machine is in good working condition before doing so.

In addition to ensuring it has all the attributes detailed above look for the following points. The obvious first thing to look at when buying a used machine is its overall condition. It is a case though of "buyer beware" as a coat of paint and a rub with metal polish can work wonders on the appearance of a machine that has in fact seen better days. So start by looking at the bare metal parts and see if there are signs

This very fine lathe was seen at an exhibition and has been home built from castings.

of wear. Pitting on the bright metal surfaces are a sign that it may not be quite as good as the paint finish infers. Chips out of the bed and cross slide casting are a bad sign and mean that the machine deserves a closer examination. Try the movement on the saddle and cross slide; see if there is any excessive play or excessive tightness. It is possible to correct a small amount of play, but a very tight movement could mean that an attempt has been made to cover up the amount of wear the machine has had. At the same time it will not be unusual for these components to be tighter at the ends of their travel than in the middle, most machines suffer in this way after a while, but neither play nor tightness must be excessive.

When you go to see the machine, take along a short length of wood, something about 2x1in and 18in long. Put one end on the bed underneath the headstock, or the chuck if it has one, and lever upwards, there should be no movement at all, if there is it means the bearings are worn and replacing them might be a difficult proposition. Move

the spindle round by hand; make sure it has no tight spots, which indicate that it may have been tightened in order to disguise wear. If the machine has a chuck, open and close it to examine it for wear. There will always be a little movement in the jaws but a couple of millimetres should be the absolute maximum and even that is somewhat excessive.

Getting the machine home

How to get the machine home will depend on what size of machine is bought. There are specialists who will transport it, ensuring at the same time it is taken care of throughout the journey. If it is proposed to transport it for oneself, remember that just because it is heavy and bulky that does not mean it cannot be damaged. If it is on a stand, and is transportable on that stand all well and good, if it is not on a stand it is best first of all to remove the motor, if possible. Motors are heavy and will put an undue strain on the machine when it is lifted. Remove the chuck if it has one. It is

The various accessories that are acquired as progress is made should be kept in a convenient place ready for use. In this case morse taper tooling is set on a sloping shelf near the lathe location. The fact that the shelf is sloping means that dirt and stray swarf either falls off or is easy to remove and also that all the tools are easily visible.

necessary to keep the machine as flat as possible and it might be a good idea to make up a wooden platform, put the lathe on it and tie it down tight. Then use the platform for all lifting operations, rather than trying to lift the bed of the machine.

Installing

The machine must be mounted either on a stand or else on a good solid bench and if possible the bench should mostly be of metal construction rather than wood. Wood twists and warps, how long it takes to do this depends on what the wood is, but it is almost inevitable that it will do so and so it is best to aim for a metal framed bench. Specially designed stands are available and they incorporate drawers and shelves to hold accessories. The bench must be level and the machine when mounted on it should also be level, if necessary level it by using washers underneath the mountings. Finally make absolutely sure that it is firmly bolted down.

Chapter 3

Power supplies

We all accept without question the use of electricity to power many of the ordinary household items and providing that we purchase a lathe that has been fitted with the correct type of motor, there is no reason why it should not be plugged into the household supply and used in that way. The idea has the advantage of being simple and convenient but is not necessarily the best solution. Not all lathes will be suitable for just plugging in and there are other matters that have to be considered, such as power consumption in relation to the maximum wattage that the household mains can accept.

With a small lathe there is not too much of a problem, motors are small enough for the household mains to cope with them, but with a larger machine matters are very different. If there were such a thing as an average lathe, then it would no doubt have a centre height of around three and a half inches or eighty millimetres. Such a machine requires a motor with a capacity of at the very least a quarter of a horsepower or 1900 watts. In order to get the best from the machine a motor of double that power would be ideal and although a little more expensive there are plenty of motors readily available. Machines that are supplied with a motor already fitted will often be fitted with a motor of larger power as standard and this could easily be three or even four times the 1900 watts. To get some idea of how motors of this power compare to those driving domestic appliances, very few vacuum cleaners are fitted with motors rated at more than 1800 watts. So a lathe motor will possibly require double the power of the vacuum cleaner. Of course household mains will cope with more than that, it is quite likely that for domestic use a washing machine, vacuum cleaner and oven will all be in use at the same time,

an arrangement whereby the motor is switched on but does not at that point actually power the lathe. Doing this involves fitting a clutch to the lathe, this will allow it to be started more gently and reduce the initial amount of current required for start up. Some people use a system whereby a drive belt is left loose until the mandrel is required to rotate and then the belt is tightened via a lever to get it started, a primitive form of clutch. Lathes can be bought with a clutch fitted and it is also possible to devise one for oneself. It is certainly a very desirable feature to have and worth considering. One thing that is to be avoided is leaving the self-acting mechanism of the carriage engaged and then applying power to the machine, this puts even more strain on the power supply.

plus possibly an electric kettle as well. The mains supply has no difficulty with what is a pretty hefty load but there is a big difference between using those machines and using a large motor for a lathe.

When we switch on and supply current to the motor, there is an additional pull of electricity to the motor. That is normal, if we let the clutch in to start a motorcar from standing, the same applies, extra power is required and we compensate with a touch on the accelerator. We do not have an accelerator to give the extra push to the electric motor and so there is a temporary increase in the amount of power taken from the main supply. In most cases it will do little more than dim the lighting for a split second, but it is possible that in some circumstances it could be sufficient to blow a fuse.

One way of easing this strain is to have

Whatever method of starting is to be used the circuit should be protected with a Residual Current Device (RCD) in Britain it became law that this should be so in 2006. Most household fuse boxes are now fitted with them and they will trip out if there is an overload of current. At least when the RCD trips out it is not necessary to hunt around for the fuse wire, as it is when a fuse blows, only to discover that the wire has disappeared from where it was last seen. Whatever the fusing arrangements for the house it is a good idea to have an RCD fitted in the workshop electrical circuit in addition to the arrangement on the household circuit. It is possible to buy small switch boxes already fitted with RCDs and putting one of those into the circuit makes a great deal of sense If it is not practical to fit one then consider fitting an RCD on to

An RCD that plugs directly into a thirteen amp socket and also accepts a standard thirteen amp plug. Useful to protect individual machines and other power consuming equipment, it can be used in addition to the RCDs in the main supply, for additional safety.

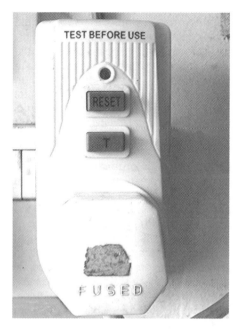

the plug itself, although not as satisfactory as one wired into the workshop circuit it is nevertheless a good safety device. Adapters that can be plugged into an ordinary socket can be obtained; the supply to the machine is then plugged into that. They look a little untidy but nevertheless do the job and a little untidiness is a small price to pay for safety. It will ensure that in the event of anything going wrong, the lathe will be disconnected from the power supply. For extra security there is no reason why both permanent and temporary RCDs should not be used.

It should be noted however that while a motor must have sufficient power to drive the machine and cope with the effort required for cutting metal, using a motor that is too powerful is not such a good idea either. These days the physical size of electric motors has reduced so much that it is tempting to go for extra power, which is fine until the unexpected happens.

This type of mains socket incorporating an RCD is ideal for the home workshop. May take time to find a suitable source of supply.

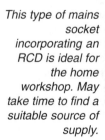

Suppose for example the tool should happen to 'dig in', the extra power will allow the motor to keep going thus pushing the tool in further, from there on untold damage could be done to the lathe. So when selecting a motor try and find the correct one for that particular lathe and if you do feel the need for a little extra power, don't go mad.

The motor should always be firmly bolted down in order to prevent any unwanted movement, most lathes will have a plate to which the motor is fixed, there are however some instances when the motor needs to be bolted to the bench or stand. Care must be taken when wiring it to ensure that the wires are held firmly in place. Most motors have a block with screw terminals but there are cases where push on connectors are used. The electrical connections, no matter what type they are must be covered, in most instances there will be a plate for this purpose, if not a suitable cover will have to be made. If it is made from metal then ensure it does not in any way come into contact with the electrical connections, in the case of push on terminals make sure that the insulation that covers the connections is in place.

Perhaps at this point it should be mentioned that in Britain it is now illegal to make ones own alterations to the electric supply, the only people allowed to do so must have a certificate of competence from the local authority. So even if you do happen to have a degree in electrical science, or have spent a whole life as an electrician, you can be prosecuted for making unauthorised alterations or additions to the household supply, unless you get a certificate from there.

The household electricity in Britain is 230 volt plus or minus 10 volts; alternating current and is similar in most countries, it is known as a single-phase supply. This means that there are two wires, brown and blue plus an earth wire. In its basic form electricity has a positive and negative side, but with a/c current this alternates fifty times per second, which means at one split second the brown is positive, the next negative and so on. Even so there is a wrong and right side to the wiring and it must be followed, and equipment should always be earthed.

Most earthing is now done via the plug; the wiring carrying back to the supply and it is there that it is earthed. Older houses may not be wired this way and can possibly have an individual earth system. It may therefore be necessary to be prepared for this as relying on the earth prong of a plug might not be sufficient.

Reversing

Whatever type of electricity is in use, it is desirable that it should be possible to run the machine in reverse, the facility will not be needed that frequently but it is invaluable. A typical example of when it might be needed is when tapping a very long hole, or running a die for a long distance on a bar. If a reverse is not fitted to the machine it becomes quite a chore winding it back by hand, not only in order to completely release from the thread, but also to clear the swarf. The temptation is to forget about winding back to clear the swarf and the result is a broken tap or worse still a broken die. Modern machines with ready fitted motors will already be capable of reversing and anyone using an inverter will also have the facility built in. Those who are buying used machines may very well find that no reverse mechanism is fitted. Most motors can be made to run in reverse, but converting a motor to do so may mean altering the way it is wired internally, in which case it is best left to someone who knows what they are doing.

Various types of reversing switch are available; the main consideration must be that it is man enough to take the current that is being used. A switch that is not rated high enough will overheat and there will be real danger of a fire, so make absolutely sure that the switch is man enough for the job

Modern electric motors are often fully encased and fan cooled, older ones have ventilation slots that are a potential source of trouble. On many machines the motor will be mounted at the back of the lathe, behind the headstock and this is just about the most vulnerable position it could possibly be in. Most of the dust and swarf generated by cutting operations, will be making its way in that general direction with the result that the motor will in no time at all be covered in the stuff as well as in any stray oil, or cutting fluids that have been in use. In addition electric motors rely on magnets to make them work and magnets will draw in fine particles of steel. All of this muck will be drawn into the ventilation slots unless care is take to prevent it from doing so.

Some machines have a guard built in to protect the motor from this hazard, if not it is essential to fit some type of guard or cover to stop it from happening. What form this takes will be entirely in the hands of the owner of the machine. It might be an overall metal casing, or could even be made from wood. One thing for sure is that any guard must have plenty of ventilation in order to prevent the motor from overheating. Following on from this is the fact that all electric wiring and fittings around the machine should be cleaned and inspected regularly to prevent any mishaps.

The on/off switch for the lathe should be mounted at some convenient point but not where it can be accidentally switched on. This applies particularly to the lever type; they can all too easily be caught by a piece of clothing and the switch moved. It must also be in a position that ensures the operator does not have to lean across close to the revolving parts in order to get to the switch and can get to it in the case of an emergency.

The electrical supply in our homes is known as single-phase. Electricity supplies to factories, large warehouses, etc have what is known as a three-phase supply, which means literally what it says. There are three wires in addition to the earth and the current switches between them alternately, and because there are three instead of two the movement of the electricity is much smoother. The voltage is also higher at four hundred and fifteen volts total. This three-phase current is used to supply motors driving machinery in factories and requires a different type of motor to those used on the single-phase supply. The result of this additional phase is to give a much smoother operation.

Anyone purchasing a used lathe that has been installed in a factory is therefore almost certainly going to get a motor designed for three-phase operation and although it is possible with a little effort and expertise to run such a motor on a single-phase supply it will not run all that well and will lack power. There are two answers for anyone purchasing such a machine, either

The ability to reverse the lathe is useful and conversion of the motor is usually fairly easy. All that is then required is a reversing switch; the photograph shows a drum switch that has a nice positive action. This type of switch should be mounted in a position where it cannot accidentally be switched on.

change the motor for a single-phase one or change the electrical supply in the workshop to three-phase. How straightforward it will be to change the motor will depend on the machine that is being purchased. If it is one with a separate motor it will be easy enough, if however it is one where the motor is more or less an integral part of the casing it could be a very difficult proposition.

It might therefore be desirable to install three-phase electricity in the workshop and the electricity supplier will be glad to do so, but invariably this is at a considerable cost. In some cases, particularly in urban areas a three-phase supply may well run right past the house. It is quite common for the supplier to use a three-phase supply and to split it between every second household. In other words they take one side to one house and the other side to the next. At

least it means that there is less upheaval involved in putting the three-phase supply in, but it is still expensive. People living in rural areas may not have this sort of supply and in that case putting in a three-phase circuit would become nearly impossible. The arrangement of dividing the three-phase supply between houses does appear at times to result in a temporary reduction in supply to one, if something that consumes a great deal of current is operating in the other. One obvious indication of this is when the power is switched on to the device, there is a dimming of the lights both in the house where the equipment is situated and the one sharing the supply.

There are other ways of getting three-phase electricity that do not involve having the road dug up and spending small fortunes. This is by using a device known

A rotary converter, suitable for converting the whole of the electrical supply to the workshop into three phase current.

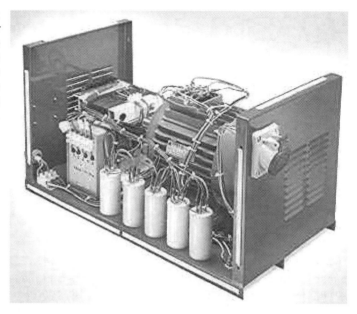

as an inverter or converter. There is a very wide range of these available, not all are designed for the purpose referred to above and those that will do the job vary considerably in the range of facilities they offer. They vary in capacity, from those that will power a complete workshop to small ones suitable for a single machine. It is also possible to get one that suits the power of the motor that is to be used with it. Most modern converters offer more than just a straight current conversion from single to three- phase and to 415volts. In addition reversing the direction of the motor is easier with three-phase electricity.

Phase converters

There are two main types of Phase Converter, static and rotary, the latter uses an electric motor, the former is a series of electrical components. The rotary converter

is particularly efficient; a decent one will be quite capable of supplying three-phase power to quite a large workshop. The use of either of these will allow the whole 415 volts of the original supply to be used, where as with an inverter only 240 volts will be available Converters are generally believed to be more efficient than inverters, but are larger and therefore less convenient to use. There is also generally a need to supply control gear for the motors where as with some inverters there is no such need. By their very nature and the fact that they use an electric motor the rotary type are noisier. Modern static units are electronic gadgets that operate quite silently and efficiently and it is this type that is becoming more and more popular in the home workshop.

Normal three-phase current is rated at 415 volts and motors for use on three-

Rotary phase converters are obtainable in a variety of sizes.

phase supplies are designed for that voltage. Most industrial motors are however rated for dual voltage and it is a simple task of undoing a couple of screws to convert one to run on 240 volts. It is usual to use induction motors with the three-phase supply and these have the advantage of being lightweight and easier to maintain.

Phase inverters

Sometimes called variable frequency drives, phase inverters are used to change the supply from single-phase to three-phase at 230 volts. With the use of a suitable switch or pod as it is usually referred to, control the speed of an electric motor, they are therefore quite suitable for use on a lathe. However most inverters are quite sensitive as regards their circuitry. They cannot be used to power all types of motor. They do have a distinct advantage in the home workshop as they can be obtained as quite small self-contained units that can be mounted on or near a machine

and used in place of the more normal electric control system and are also relatively cheap. The inverter differ s from a phase converter in as much as it allows the speed of the motor to be controlled via a suitable switching arrangement, whereas the converter supplies a regular current that is not so controllable.

There is one other big advantage to using three-phase power and it is that the lathe will run a great deal smoother than it will with single-phase electricity. With single-phase there are split seconds of hesitation in the current flow. They are so short that under normal conditions they are not noticeable, but they are sufficient to ensure that the finish on a piece of metal will not be perfect. All lathes, or come to that any metal working machinery has the ability to flex, even if under normal circumstances it is not noticeable. The tiniest stop/start will be sufficient for the machine to do so, thus causing the tool to dig in and give a tiny unevenness to the finish.

A small inverter suitable for use with a single machine.

The use of three-phase electricity is strongly advocated, but readers are advised to get professional assistance before deciding on the type of converter to fit.

If a conversion is being made there is no point in fitting the motor in such a way that it drives the countershaft. The main reason for the countershaft is to allow the belt from the cone pulley to drive the cone pulley on the lathe mandrel. The cone pulley is made obsolete as is the back gear arrangement as all speeds can be controlled via the inverter. If possible then the motor should drive directly on a pulley on the lathe mandrel, the countershaft is just absorbing power unnecessarily. This is not always possible and so it may be necessary to use the countershaft. An example of this is with some Myford lathes, to drive direct to the mandrel involves removing the countershaft and a bracket that holds the motor plate, otherwise a belt cannot be fitted. It is possible to build another bracket and support it using the tapped holes in the back of the lathe. It means also dispensing with the cover for the original drive belt and in the interest of safety new covers will have to be made.

There may also be a problem with lathes that have a clutch fitted as that is nearly always incorporated in the countershaft. If it is thought necessary to retain the clutch then the drive will have to be on the countershaft. One other advantage of using a three-phase motor is that it will be possible to increase the speed at which the mandrel will rotate. This means that there can be a so-called soft start from zero, which in many ways makes the clutch obsolete. Older lathes have what is now considered to be a comparatively low top speed. This applies particularly if tipped tooling is to be used, where faster turning rates are obtained. Some thought should therefore be given to the size of belt pulley that is fitted to the motor, when converting to three-phase. It should be possible for the top speed of the lathe to be increased by at least fifty percent. Fitting a larger pulley wheel to the drive shaft of the motor can do this. It is not advisable to increase the lathe speed too much as in all probability the bearings have been designed to work at a set range of speeds. There will be no difficulty at the lower speed range as that can easily be achieved by using the speed control.

If a lathe is converted to run on three-phase current it means that the speeds that are normally given by the changing of puller permutations no longer work. It is therefore advised that a tachometer should be used to check the speed at which the lathe is running. Tachometers can be obtained that work either by friction or by electronic

37

A control box or pod to control a motor used in conjunction with an inverter

indication from a sensor. They can be machine mounted or hand held and any type will be suitable as long as it gives a correct indication of the speed.

Lighting

Good lighting on the machine is essential and while the best possible form is daylight some electric lighting will inevitably also be required. Good general illumination helps and a movable lamp that can be directed on to the working area is even more important. The wattage of the bulb or bulbs in this is a matter of personal choice, ordinary tungsten bulbs get very hot and sixty watts is possibly about the maximum that is practical. All bulbs of this type are due to become illegal, which means an alternative form of lighting will be necessary anyway. The replacement for the tungsten bulbs are the low energy bulbs called compact fluorescent lamps or CFLs, and while they might be suitable for general household purposes they are not ideal for illuminating workshop machinery. They take considerable time to warm up and reach full lighting capacity and are generally too large for the job. Another problem with them has been reported by some people, with all fluorescents there is a flicker that coincides with the mains power at fifty cycles per second. It has been suggested that when machines run at certain speeds there is a stroboscopic effect that would make the machine appear stationery, thus creating an additional danger.

The most practical idea is to use a low voltage lamp, with its transformer situated some where away from the general working area. Mains voltage lighting close to machinery is always a potential source of danger and should be avoided. There is always the danger of accidentally breaking the lamp and if it is mains operated this could result in an electric shock. It is also possible that the vibration of the machine could wear away the cable insulation and the whole machine become live. For safety using low voltage lighting makes sense. At one time this meant using bulbs that gave of a very dull light. It is now possible to use halogen bulbs that have the advantage of

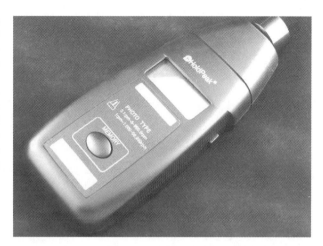

Having converted to electronic speed control it is no longer possible to use the variation of pulleys to obtain the lathe speed. It therefore is advisable to use a tachometer. The photograph shows a hand held version, it is possible to obtain types that can be permanently mounted as well.

being small and powerful with a twenty-watt bulb being more than sufficient to illuminate the working area of a machine. If they suffer from any disadvantage it is the fact that they get very hot and accidentally brushing ones unprotected hand or arm against them can be quite painful. A decent cover or shade will prevent this from happening.

Arguably the best solution of all is to use the modern Light Emitting Diode (LED) type of bulbs. LEDs have been around for many years but it took some while before the idea of grouping a lot together to create a single lamp came forth. They are now easily available, as low voltage units and are far cheaper to run than the CFL and very much brighter as well. They do not suffer from the flicker effect of the CFL either. Many LED lamps offer a choice of brightness; a certain number of units are brought into play for the lower volume light and switching in additional ones increases the brilliance. LEDs are supposed to have an almost indeterminate life and in addition they produce very little heat, making them ideal for illumination the work on the lathe.

At all times remember that electric current can be lethal so treat it with care.

Chapter 4

Cutting tools

Although a variety of tools will ultimately be used on the lathe, the term lathe tool is generally applied to mean those intended for turning and boring operations and the first to be considered is the turning tool. At one time the material used to make them was high carbon steel, as the name implies this is steel containing a high quantity of carbon. Tools would be filed to shape from the material and then hardened; they would then be tempered, subsequently being honed to a fine edge. It is a process rarely used these days but is still a very good way of making specially shaped tools and will be covered at greater length when we come to form tools.

The introduction of high-speed steel gave the operator the opportunity to make tools that would last and retain their edge considerably longer than those from carbon steel and it rapidly became the number one material to use. Now it was a case of grinding a short length of high-speed steel to shape and this is still a practice carried

Below Left: A general-purpose tool in a cranked holder. This type of tool and holder is particularly useful when a single clamp is used to hold it. Although the tool bit is removable, generally speaking it is advisable to retain it in the holder when sharpening it. Below Right:This tool holder is a comparatively modern design and the shape allows the machining operation to be clearly visible.

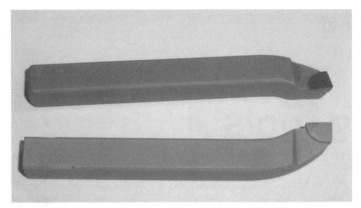

A pair of tools with cemented carbide tips.

out by many. Even harder and tougher materials such as tungsten carbide followed high-speed steel and tools made from this are available and widely used. This is not true steel being made of various alloys in powder form that are fused together and moulded to the required shape. It is mainly used for lathe tools in two forms, either tips that are brazed to the tool, or replaceable tips that are either screwed or clamped to the tool. Because the material is so hard it is also brittle and so it is necessary that the tool shank provide plenty of support. It is possible to buy short pieces of the material and to grind it to shape for oneself, but this is not a common practice. The advent of high-speed steel did not completely sound the death knell for high carbon steel, which still has its uses even today. Likewise tungsten carbide and similar materials have not entirely superseded high-speed steel. All these materials have their uses although in general the purposes are very different.

The angles at which tools are ground, or in the case of carbide moulded, have been developed over the years and no apologies are made for reproducing them

once again here. They have proved the most suitable for the majority of machining operations and have stood the test of time. They are not however statutory and many operators find that different angles are better suited to their work, even so the standard ones should form a basis on which to work. A fairly common sort of variation is with the front rake, which is there for both clearance and strength. When doing heavy work it is essential that the front edge be well supported and that is the reason that the angle of around ten degrees is suggested. It works perfectly on steel and large diameters, but is not quite so suitable for softer material, particularly where very tiny diameters are being machined. It will therefore often be found to be an advantage to increase this angle and it can also be found to be beneficial to do the same with the side rake. Of course this can only be done in the case of tools made from carbon or high-speed steel, hard tips are not suitable for modification. Therefore while the correct angles should always be kept in mind, different angles and perhaps even different shapes may well be better suited to different conditions. The best results will

TOOL CUTTING ANGLES

Three figures are given for each type of material. The first is for roughing out, the second for fine finishing and the third for parting off.

Material	Back Rake	Side Rake	Front Clearance	Side Clearance
Mild steel	6-10	16	5-9	5-9
	14-22	Zero	5-9	1-3
	15	Zero	5-9	1-2
**********	**********	**********	**********	**********
Silver Steel	6	12	5-9	5-9
	10-12	Zero	5-9	5-9
	5-10	Zero	5-9	1-2
**********	**********	**********	**********	**********
Cast Iron	8	12	5-9	5-9
	6-10	Zero	5-9	1-3
	6	Zero	5-9	1-3
**********	**********	**********	**********	**********
Brass	Zero	Zero	6	5
&	Zero	Zero	6	6
Gunmetal	Zero	Zero	6-10	2-3
**********	**********	**********	**********	**********
Copper &	8-18	16-25	5-9	5-9
Phosphor	10-20	16-25	5-9	1-3
Bronze	10	Zero	5-9	Zero
**********	**********	**********	**********	**********
Aluminium &	8	15-22	5-9	6-10
Similar Alloys	8	15-22	5-9	1-2
	10	Zero	5-9	1-2

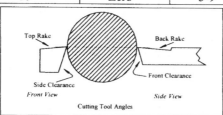

Cutting Tool Angles

A tool with a replaceable tip that is screwed in position. Note that the tip has three cutting edges, allowing it to be indexable and therefore can be rotated to another cutting edge if it gets blunt or damaged.

of course only come as the result of experience.

Tools with brazed on tips are for some reason described in catalogues as cemented, or sometimes as butt-welded no doubt some tips are retained with one of the modern high strength adhesives or cements, most are brazed. Many if not all manufacturers of these tools insert a piece of copper shim between the tip and the tool, to act as a buffer should the tool receive a heavy shock. It is only inserted at the lip where the tip butts against the metal. Tipped tools come in a very wide variety of sizes and shapes and are frequently sold as complete sets. Unless it is financially prudent to do otherwise the best idea is to decide the types likely to be required and just get those; it is always possible to add to ones collection, in the light of experience.

Sharpening tools with cemented tips is a slight but not insurmountable problem; the difficulty is that a normal grinding wheel will not work on the tip. This is because grinding wheels are made of millions of tiny pieces of sharp abrasives all stuck together. The hardness of the tool blunts these almost immediately they touch and

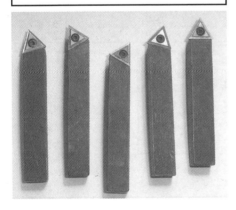

Copper shim is cemented between tip and shank to assist in prevention of tool chipping

A set of tools with replaceable tips that has been designed so that even though the same type of tip is used for each tool a wide variety of operations is possible.

therefore they can no longer do their job. Wheels are available that are designed especially for the work, they are generally called green grit wheels, actually it is the bonding that is green not the grit. This bonding wears away rapidly taking the blunt abrasive with it and leaving sharp pieces that are able to deal with the tip. A more modern method is to use a diamond disk, as that will do the job in no time. The disks are available in a variety of sizes. Anyone that does not have the facilities to mount such a disk might like to try one of the mini drills with a small diamond disk fitted. The turning tool can be held firmly in a vice while the disk is used to grind it, The idea works quite well and although a new diamond disk may be necessary for each tool that is being sharpened they are ridiculously cheap to buy.

All tools that have been ground should be honed afterwards, in the case of high-speed steel oilstones will be sufficient but cemented tipped tools must be honed with a diamond lap. There is of course no reason why a tool made from high-speed steel should not also be honed with the diamond lap, but don't try and hone a tipped tool with an oil stone as it will be a waste of time.

People often ask why on earth we should hone a tool when it has just been ground to a keen edge? And the answer is in the above paragraph. The grinding process consists of rubbing lots of tiny sharp particles against the tool and as these are particles, the edge that is left is very ragged. It is not noticeable to the naked eye but put under a magnifying glass it can be seen very clearly. A ragged edge will mean a ragged cut and therefore a ragged finish, honing the tools removes the raggedness allowing a much better finish to be obtained.

Indexable or throwaway tips on tools are very effective and rightly popular, a very big advantage is should the cutting edge break or wear so that it is no longer efficient

Just a few of the very wide variety of indexable tips that are available.

45

A simple form tool for making a small external radius, note that a radius has been formed on each end.

of swarf and less heating up of the tool. The coating is very thin; in fact it is measured in microns and therefore will only apply to a cutting edge when the tool is new. Any sharpening process will remove the TiN coating from the cutting edge itself, but it will still be effective on secondary surfaces.

Certain tooling is available in solid Tungsten Carbide, drills and centre drills are those that are most seen but it is also possible to obtain parting off blades made from the same material. Sharpening of these follows the same methods described for cemented tips. It is possible to buy other hard materials in a similar form. These include Carbide, Cobalt and Stellite. Cobalt in its metallic form is harder than high-speed steel but can be ground to suitable shapes for operations on the lathe. In recent years it has lost favour to carbide, which is tougher still. Stellite is an alloy of Cobalt, Chromium and Tungsten. Tungsten Carbide incidentally has Cobalt as one of the alloys from which it is made.

Care must be taken when using any of the above alloys and this of course includes the tipped tools referred to earlier. While they all will give excellent machining properties none are suitable for making intermittent cuts as they do not withstand shocks very well. Therefore if machining uneven castings or square bar stick to high-speed steel tooling as Carbide, etc will almost certainly break or chip under such treatment,

it is a simple job to replace the tip without changing the tool setting in any way. There are now plenty of these tools available designed for a wide range of purposes and it is even possible to get some that are designed for parting off. Tips are secured either by holding them in place with a screw, or in the place of larger ones by clamping them to the body.

TiN coating

Readers will no doubt have seen throwaway tips that are a bright gold colour, this does not mean they are made of gold but that they have been coated with TiN, not tin as used in food cans but Titanium Nitride. Drills, taps, milling cutters, etc. can all be found that have been coated in the same way. Basically it is a form of ceramic coating that provides a very hard outer shell to tools; this in turn means a quicker release

Form tools

Form tools are tools that are shaped especially in order to machine something

A pair of form tools specially shaped for a particular purpose. The shapes were filed out and the tool hardened and tempered.

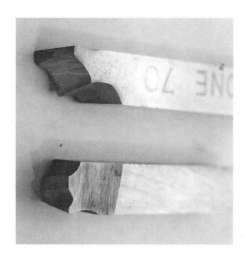

to a particular design. It might be a round section, internal or external, or perhaps a very short tapered section that is difficult if not impossible to make in any other way. The easiest and most usual way of making these tools is to file them to shape and then harden and temper them in the old fashioned way. The two main types of high carbon steel that are readily available are silver steel and ground flat stock, otherwise known as Gauge Plate. Silver steel is generally only available in round sections, and Gauge Plate can be obtained in square, rectangular and flat. When filing the tool to shape remember that it is necessary to have a cutting angle if it is to be successful. If silver steel is used it will of course be necessary to make a holder for the finished tool and the top of the bar must be removed to leave a flat surface. If the tool that is wanted is for the purpose of making a simple external radius, this can be done by simply drilling a hole of the

appropriate size in a piece of suitable steel and then filing away the end and edge to produce the finished tool.

Both the above steels are treated identically when it comes to hardening and tempering, they are heated to a bright red colour and quickly quenched. They then need to be cleaned until the original brightness is restored and heated until they turn dark brown and are then quickly

A small tool, in this case a taper reamer being tempered in a tin of sand.

A high speed tool in use, the clearance angles can be clearly seen in this photograph.

quenched again. The recommended medium for quenching is whale oil, something that will be impossible to find in this day and age. Any good vegetable oil will do instead but take great care as hot oil will almost inevitable spurt out when the hot metal is dropped and also is likely to catch fire. So do not use a plastic container to hold the oil. It is as well to have some sort of metal cover handy that can be quickly put over the container should this occur, whatever happens if it does catch fire do not put water on it.

Cleaning the metal after the hardening process and before tempering can be very difficult, the only way is plenty of hard rubbing with emery cloth. The task can be eased considerably by coating the work in washing up liquid before heating it, the liquid forms a sort of coating that cleans off rather easier and it does not stop one seeing when the metal is hot. Tempering can be done in a couple of ways. The opposite end to the cutting edge can be

heated and the metal will be seen to be gradually changing colour away from the heat, the moment the cutting edge reaches dark brown quench it.

An alternative and much better way of tempering is to use a bed of sand, it does take considerably longer than allowing the colour to run but it is possible to exercise a greater degree of control. All that is needed is a small flat tin, an empty sardine tin is ideal, half fill it with fine sand, silver sand or something similar will do, the exact type is not important. Prop the tin up on a couple of bricks, one at each end and apply a blowlamp to the middle. The tool will be seen to slowly change colour and will allow more time to quench it than the previous method.

Should there be any doubts about using oil to quench the tool, it is possible to do so with cold water. Using water there is always a slight danger that the tool will crack as it is plunged in but that is not inevitable and readers may prefer to do

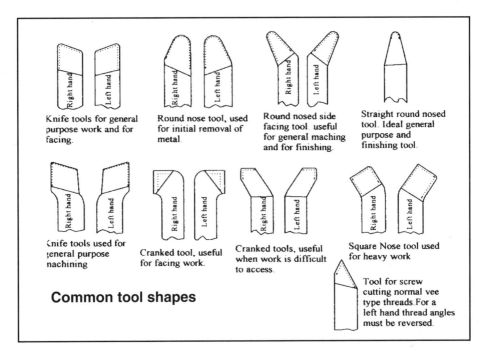

Knife tools for general purpose work and for facing.

Round nose tool, used for initial removal of metal.

Round nosed side facing tool. useful for general maching and for finishing.

Straight round nosed tool. Ideal general purpose and finishing tool.

Knife tools used for general purpose machining

Cranked tool, useful for facing work.

Cranked tools, useful when work is difficult to access.

Square Nose tool used for heavy work

Tool for screw cutting normal vee type threads. For a left hand thread angles must be reversed.

Common tool shapes

things that way rather than take a chance and use oil.

Readers will find out for themselves the tools that suit them, it is a case of what suits one person may not suit another. For the beginner the best way is to start with tools ground to the normal standard and experiment a little to see what suits one best. Try high speed and tipped tools and again see which type is best suited to your work. Carbide tools of any type are designed for machining at high speed. If the lathe does not have a very high top speed it could be that they will not work as well as high-speed steel, in which case don't use carbide tools just because they are the most modern. Use what suits you and having made your selection stick with it, the best results come from tools that one is comfortable with.

Chapter 5

Work holding - chucks

Holding work is of course one of the most important factors to be considered and various suggestions are given below. The most important thing to remember whatever method is being used is that the work must be firmly held and absolutely secure. A thorough check to ensure that everything is right should be made before starting any machining operations. It is not unknown for work to suddenly come off the lathe during operations and while unless the machine is operating at very high speed it is unusual for it to be thrown across the workshop, a heavy piece of metal falling on one's foot is not a laughing matter.

For the sake of convenience all work holding devices when discussing the method of securing them to the machine spindle will be referred to as chucks. There are a variety of ways with which they are likely to be attached to the lathe spindle, depending on the make of the machine and to some extent on is size. One of the most common methods is for the end of the spindle to be threaded and the device to be screwed in place. There is usually a collar on the spindle, the chuck is screwed

A common way of securing a chuck to the lathe mandrel is to screw it on; the photograph shows the rear of a four-jaw chuck that screws directly to the lathe.

The rear of a chuck showing how it is fitted by means of a tapered recess, that locates with an identical taper on the lathe mandrel, it is then locked in position by twisting the three cams.

fitted in a home workshop, doing so also requires making the spigots. The lathe mandrel will have a shallow taper and a matching internal one will be on the lathe back plate.

There are also examples of chucks fitted with bolts on larger machines and a taper is used to assist in mounting the chuck. Chucks are heavy items and the larger the lathe the heavier they get, some sort of support underneath them when fitting and dismantling is a necessity.

Fitting a back plate

For some lathes with screw on fittings it is possible to purchase chucks that are threaded ready for use, in many cases this is not so and they have to be fitted via a back plate. This is a round plate, usually of cast iron but sometimes of steel that bolts to the rear of the chuck. In order to mount the chuck as accurately as possible it is usual for the operator to machine the back plate him or herself. This may involve machining the matching thread for that particular lathe although for popular machines it is possible to purchase a ready threaded version.

If the plate has not been threaded that will be the first job and in order to do so it will either have to be mounted in a four-jaw chuck or on a faceplate, both of which are dealt with later. The centre will have to be bored to the core size of the required thread and then it can be screw cut to fit the lathe. Screw cutting is also dealt with at length

right up to it and this gives a simple form of locking. It is essential that the chuck or other device is screwed up as far as possible, failure to do so can result it moving while machining operations are carried out. When the machine is rotating in reverse, notice should be taken of any tendency for the chuck to unscrew. Providing it has been well screwed on in the first place there should not be a problem, but there could be a danger of it happening if the thread of either part happens to be worn, or badly fitting for any other reason.

A method used for some small machines is for the chuck to be located on a spigot at the end of the spindle and secured to it with screws that pass through the chuck body, although this method can only be used with certain types of chuck.

Another common and popular method is known as the cam lock system, the chuck is located on a spigot that is part of the actual mandrel and is secured with cams that provide a very firm locking system. The chuck does have a back plate but it is unusual but not unknown for these to be

Most screw on type chucks first require to be fitted to a back plate, this is because the threads vary according to the manufacture of the lathe. The photographs show the rear of such a chuck and its back plate. Note how the back plate has been machined to an accurate fit in the recess in the chuck.

later in this book.

Assuming the plate has been threaded it must be screwed on the mandrel and the first operation will be to machine across the back, for that it will be necessary to fit the plate the opposite way round to which it will ultimately be used. It can then be turned round for the front also to be faced at which stage it should be true both sides. The next job will be to machine the spigot that fits in the recess in the chuck and this needs great care, as the accuracy of the chuck will depend on it being a good fit. The best way to do it is to make the spigot oversize by about 1/16in or 1.5mm and about 1/4in or 6mm too long. Machine a length of about 1/8in or 3mm to what is believed to be the exact diameter required, if in any doubt make it the tiniest bit oversize. Offer the chuck up to it and see whether it does fit, if not keep alternately machining a little from the diameter, until a suitable fit is obtained. This has to be such that that chuck needs to be pushed hard to get it on. It is essential

that when doing this that the chuck is absolutely parallel with the spigot, if it is not, the fit will not be right. A useful idea is to set the chuck up on wooden blocks to the required height, otherwise it is all too easy to tip it to an angle, and not realising one has done so. Once certain the required diameter has been reached machine to that diameter for another 1/8in or 3mm and try the chuck once more. If the fit is right it can be faced across until the lip is the exact length of the recess in the chuck. Remember the accuracy of the chuck will depend on the accuracy of the fitting of the back plate.

The holes to take the securing bolts should be the next job and they should be drilled and reamed to size but first need to be accurately marked out. Start by coating the rear of the casting with marking fluid, cast iron is not the best of material for showing up the scribing marks that will need to be made. The diameter is required first and this can be obtained by measuring with odd leg callipers and again care is needed

 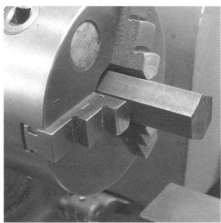

The three -jaw chuck basically will only hold round or hexagon bar and although it is quick to use is severely limited by that fact.

as accuracy is absolutely essential, once the hole diameter has been established scribe a circle. To mark off the holes measure the diameter of the circle and divide by the required number, it is in almost every case three. Set a pair of dividers to that measurement, make a centre punch mark on the circle and mark off the other two divisions. Do not rely on these being right, put one end of the dividers against one of the other marks and check that the opposite end accurately dissects the original marks. If not adjust the dividers to the correct distance, only when absolutely certain of the measurement should the two further centre punch marks be made. Take the casting to the drilling machine, ensure that it lies at ninety degrees to the machine mandrel, start the holes with a centre drill, then drill and ream.

It is essential that the screws are a perfect fit and so it is a good idea to drill them undersize to start with. Offer the back plate up to the chuck and it will be possible

to see how accurate they really are if all is well open out them to their full size and counter bore if it is proposed to recess the heads on the fixing screws. Hexagon cap screws are possibly the best type of fixing to use for this job. If by some chance the holes do not line up perfectly with those in the chuck, use a round or half-round needle file to widen them a little in the direction required. Then use a taper reamer to open them, all being well the reamer will follow the line of the filing. Even now do not open them to full size, but take another look and repeat the operation until they really do line up. Hopefully all this will not be necessary but it is not unknown for things to go wrong and at least there is a possibility of retrieval if care is taken

The three-jaw chuck

The most common work holding device is the three- jaw self-centring chuck, it is easy to use as all one has to do is insert the bar of metal and use the chuck key in one

A three-jaw chuck should have outside jaws like these supplied, they are used for holding large diameter work.

location to tighten it. It will have two sets of jaws, one for holding work between them, the other to hold it on the outside of the jaws. The big snag with this type of chuck is that basically the only metal sections that can be held in it are round and hexagon, inside and tubing outside, which severely limits its use, although it is possible to make collets that will hold square material. In addition it is very rare to find a three-jaw chuck that is really accurate, no matter how carefully the back plate has been fitted, sooner or later it will run out of true. Assuming the amount of run out is not too excessive it is best to leave well alone and when working make adjustments that allow for the run out. The eccentricity can be for a variety of reasons, a minute amount of wear on the jaws, or perhaps the jaws have been strained by over tightening, it can even be a build up of swarf and dust in the scroll. Sometimes it is indicative of a worn scroll. Overstraining is usually the result of

malpractice when tightening up. It is not unknown for people to put a length of tubing on the handle of the chuck key in order to apply extra pressure, or even to use a hammer for the same purpose. The manufacturers know how much pressure can be applied to their chucks and supply the length of handle necessary to obtain it, trying to increase it is asking for trouble.

The problem of swarf and dirt is of course easily cured, strip the chuck down and thoroughly clean it and everything should be all right. Muck can also work its way in behind the back plate and create problems, so it is worth removing that and cleaning there. It is a task that should anyway be carried out at frequent intervals in order to keep it in good condition and while it is in pieces some light oil should be applied to the scroll. As well as cleaning the jaws and scroll, make sure that the fittings on the lathe as well as the mating ones on the chuck are absolutely clean, all

55

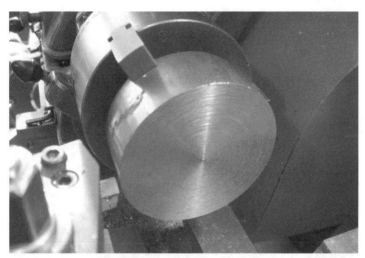

Outside jaws of a three -aw chuck in use.

these things when dirty can cause inaccuracy. Before doing so make a mark on both back plate and chuck to ensure they are reassembled in the same order. Putting them back the wrong way round can also be a reason for eccentricity. However it might sometimes be something we can use to advantage, if a used machine has been purchased and the chuck is running out of true, it may be worth while fitting the back plate, using different holes, in case somebody has assembled it wrongly.

If the jaws have become worn they may have to be ground, in order to correct things, but this should only be done as a last resort. If the problem is caused by a worn scroll, then often grinding the jaws only moves the problem form one diameter to another. To grind them a tool post grinder is needed, this is a grinding head driven by a motor, with a means of fitting it to the lathe top slide. It is hardly worthwhile dashing out and buying one, not only are they expensive but it will spend most of its time underneath the bench doing nothing. Hopefully if the

machine is used with care jaw grinding will be an operation that only needs to be carried out once. Adapting one of the small mini drills that are available can make a simple device, that will act as a tool post grinder, some of these are mains powered, others low voltage machines, it matters not which is used. Simply devise a method of holding it horizontally on the lathe, fit it with a grinding wheel and it will do the job

To actually grind the jaws it is essential that they be in tension when the job is done, that means that the top of the scroll must be in contact with the mating sections on the jaws. The time honoured way of getting this right is to put a washer at the rear of the jaws and tighten them on it, the jaws can be ground right up to the face of the washer, which is then removed and placed at the front of the jaws. The grinding wheel is passed through the washer and the last bit of the jaws trued, using the same cross slide setting as for the first grinding operation.

When the jaws are only a tiny bit out of

A small drill mounted in the tool post for truing the chuck jaws. This operation should only be carried out as a last resort if the jaws are badly out of true.

true and if satisfied that grinding is essential it is possible to simplify the job. Take a piece of brass rod, of the maximum diameter that can be held in the tailstock drilling chuck and long enough to fit in the main chuck jaws and pass right through them. Coat the brass with a grinding paste of the type used to grind in the valves on motorcar engines. Close the chuck jaws but do not tighten them sufficient to grip. In other words all three jaws must be in contact with the brass, but not closed sufficiently to actually hold it. Start the lathe and let it run at a slow speed for about ten minutes, then adjust the jaws so that the same tension is on as before and run the machine again. It is not possible to say how much grinding is actually needed, but it will be possible to true the jaws in this way.

After grinding the jaws, whichever method is used it is essential to remove all traces of the grinding process. The residue left by grinding or lapping is likely to work its way into the chuck scroll, or indeed to other parts of the machine and will rapidly

cause wear.

The four-jaw independent chuck

Many people, particularly those with little experience of using a lathe, fight shy of using the four-jaw independent type of chuck and yet it is the most accurate way possible of setting up work. As the name implies it has four jaws and they are all adjusted independently, It has many advantages over the three-jaw self centring type as not only will it deal with the same sort of work as that does, but in addition will hold square, rectangular and even odd shaped material all of which can be set to run absolutely true.

It is setting up the work in the four-jaw chuck that creates most problems for people and one of the main causes is to try and adjust work from both front and back. Decide which side you want to adjust from and sticking to it is far easier than trying to work from both. Use a scribing block to gauge the position of the work and put a piece of white paper or cloth underneath in

57

A piece of square bar being prepared in the four-jaw chuck. Because the jaws work independently it is necessary for the operator to manually set the work in the correct position. In this case we see it being done be using a wobbler or wiggler and a clock gauge. The wobbler is a spring loaded device, one end of which goes in the tailstock centre, the other in a centre mark in the work. The chuck is rotated by hand and any movement that is not concentric is adjusted by altering the position of the chuck jaws. The accuracy in this case is being set with a clock gauge, an instrument that has a spring probe to rest on the wobbler, any eccentric movement being registered on the clock face.

such a way that it highlights the gap between block and work. Do not tighten the jaws too hard when adjusting work, it is necessary for it to be capable of sliding across them while adjustments are made. When the work has been finally adjusted go round each jaw, tightening by a quarter turn of the chuck key, then go round again and repeat the operation until the work is held tightly. Tightening one jaw hard and then moving to the next is likely to move the work out of true. When setting up round bar in the four-jaw chuck, final adjustments should be made using a clock gauge,

It is possible to hold odd shaped castings in the four-jaw chuck by means of using inside and outside jaw fittings. Unlike the three-jaw self-centring chuck, two sets of jaws are not normally supplied with this type; instead it is possible to reverse them. In

addition to holding odd shapes the chuck can be used to set up work eccentrically in order to either bore it or machine the outer surface. The centre of the required bore or circumference has to be marked and centre punched, the work is then set up using a centring device, commonly called either a wobbler or wiggler. This is done by putting the point in the centre punch mark and the other end, which has a small hole in it in a centre in the tailstock. Adjustments can be initially made with a surface gauge; final adjustments should always be made using a clock gauge, the probe of which should be located on the wobbler as close to the work as possible.

In addition to the two main types of chuck detailed above there are several other types available as follows. Four-jaw self-centring, similar to the three-jaw self-centring but with an extra jaw and has little

use except for holding square bar. Six jaw self centring chuck is used for holding round and hexagon bar stock, mainly used in industry as this type of chuck generally has greater accuracy and better holding power than the three-jaw. The two-jaw chuck is also used mainly in industry for special purposes and is not a great deal of use in the home workshop. Frequently the jaws on such a chuck are soft so that they can be machined to accept specific work.

Soft jaws

The jaws on chucks are hardened to prevent undue wear and thus loss of

accuracy, this can result in delicate work being marked or even having slight indentations. Soft jaws are available that can be machined to accept particular types of work such as thin disks. They are also useful where absolute accuracy is essential, as it is possible to machine the

Another view of work being set in the four-jaw chuck, using a scribing block, A piece of white paper has been laid on the bed of the lathe to make it easier to discern any gaps left where the scriber is not in contact with the probe of the wobbler. Note that for this operation the jaws of the chuck have been reversed unlike the three jaw which requires two sets of jaws the four-jaw only needs one. It is possible to use each jaw either way round which is an aid when setting up awkward shapes.

jaws before setting the work. The same principle of ensuring the scroll is in contact is applied to machining soft jaws as to grinding ordinary jaws. Generally they are a set of ordinary jaws with attachments that are screwed on and it is the attachments that are machined prior to working. Different manufacturers use different types of attachments, some are rectangular, some round and others hexagon shape, The shape is not important as they are to be machined and its only purpose is to allow the attachment to be rotated so that several machining operations can be performed.

Collets

The most accurate way to machine round material is to use a collet, Most require a special holder which is like a mini chuck. Different types of collets require different chucks or holders and in addition a different type is likely to be needed according to which make of lathe they are required for. There are a wide variety of collet types and which to use is really a matter of personal preference. For simplicity it is possible to use an ordinary split collet with a suitable morse taper and close it by using a threaded draw bar from the rear of the lathe mandrel. Apart

Two-jaw chucks are also available, but unless required for some special purpose have little practical use.

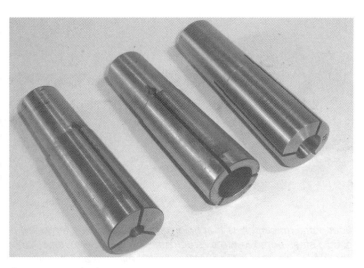

The most accurate way to hold round material is with a collet of which there are a variety of types available, the picture shows the basic form of morse taper collet that is closed by a draw bar. Other types are closed by means of caps or chucks fitted to the lathe mandrel.

from the fact that they impart a good grip to the work, without marking it, collets have the advantage of being very rigid when holding the work, because there is little overhang. In addition to collets for round work it is possible to get them for holding square bar as well, this is particularly useful as setting up square bar accurately can be time consuming, it is much easier to slip it in a collet and tighten up.

Tailstock chuck

The tailstock chuck as the name implies is intended to fit in the tailstock in order to hold drills, etc and is probably the most necessary attachment and also the most used one. As well as being used in the tailstock there is no reason whatever

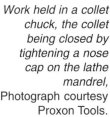

Work held in a collet chuck, the collet being closed by tightening a nose cap on the lathe mandrel, Photograph courtesy Proxon Tools.

Four home made collets that hold square material in a three-jaw chuck. They are made of mild steel, bored to the size across corners of the material for which they are intended. A saw cut is then made along the length and this allows the collet to close under the pressure of the chuck jaws closing as can be seen in the second picture.

that one cannot be held in the mandrel where it is capable of holding small diameter work for machining. Depending on the quality of the chuck it is likely to be a more accurate way of holding small diameter stock than the three-jaw chuck.

Most chucks of this type are closed and tightened with a chuck key, recent years however have seen an increase in popularity of keyless chucks that are tightened simply by hand held pressure. As with all things a great deal depends on the quality, cheap keyless chucks tend to lack accuracy as well as being difficult to tighten, better quality ones are as accurate as those closed with a key, but how much they can be tightened does depend to some extent on the strength of the operator.

The jaws of chucks are hardened and in certain circumstances can cause damage to the work. To prevent this it is wise to use soft jaws and the picture shows a typical set consisting of hexagon material. Suitable recesses to accept the work to be done is machined in the hexagon. The jaws can be rotated as they wear and are easily replaced when too badly worn for further use.

On most lathes the tailstock chuck can be fitted into the mandrel and is useful for holding small diameter material. The picture shows one being used for that purpose.

Chapter 6

Other work holding methods - the faceplate

In many ways the faceplate is the most versatile of all methods of holding work, although it does suffer from the problem that it is time consuming to set up. So far work holding has dealt mainly although not entirely with regular shaped items, such as bar stock, very often life is not as easy as that and we need to machine odd shapes, in particular this applies to castings.

Although sometimes these can be held in a four-jaw chuck, more often than not they will need to be mounted on the faceplate, one of the most useful lathe appliances. As readers will no doubt know, the faceplate is a round flat plate that fits on the mandrel and contains several slots. Because it is a comparatively thin item and usually of cast iron, the rear will have some strengthening struts. These are a necessary evil but do from time to time get in the way when work is being mounted on the faceplate. Work is generally bolted either directly or indirectly

A typical faceplate, in this case it is seen mounted on an assembly mandrel. Most faceplates follow this sort of patter, a plain cast disk with strengthening webs at the back, They size and shape of the slots varies according to the manufacturer, this example is from Myford Ltd.

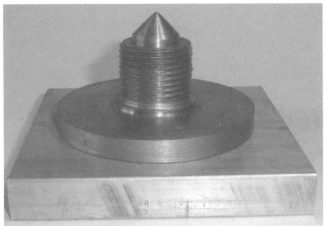

to the faceplate and that is of course the reason for the slots. It is tempting to use ordinary nuts and bolts for securing the work but it is advisable to use tee nuts as the hexagon head of a normal bolt puts a strain on a small area and in certain circumstances this could cause a breakage of the casting around the edge of the slots. It is actually worthwhile spending a few minutes to make a few tee bolts especially

*Left:Work being assembled on the faceplate, away from the lathe, Once it is securely bolted in place the whole set up is taken to the lathe and fitted on the mandrel ready for use.**Above:**Work is usually fitted to the faceplate with tee bolts like these. Ordinary bolts should never be used for the purpose.*

An alternative to tee bolts is tee nuts. Similar in design, except that a bolt is screwed into the nut instead of the nut being screwed on to the bolt.

to fit your own faceplate. An alternative is to make tee nuts, these are the same shape as tee bolts but do not include the threaded rod, lengths of studding are kept on hand to screw into the nuts when required.

Whenever a clamp is used it should always have an angle towards the work piece that is being secured, if this is not done the piece will not be held securely. Clamps should be slotted in order to allow them to have that angle, if facilities are available to do so they can be bent over at one end to facilitate this. In many instances it will be necessary to put packing under the opposite end to that holding the work, in order to get this angle and rather than scratch around trying to find a piece of packing when doing the job it is wise to have packing pieces ready for use. Every effort should always be made to ensure work is secured in at least three places, unless the shape of the object makes this impossible.

Sometimes it is not practical to secure the work with clamps alone and in that case it may be necessary to use an angle block, or even pieces of metal to give extra support. It is impossible to give a description of every set-up as work will differ according to the projects one is making. The important thing is that anything mounted on the faceplate must be held as securely as possible.

Good stout clamps should be used for securing work and the end resting on the work should be a slight downward angle, These commercially made clamps have a section at ninety degrees for this purpose. There is no reason why the clamps should not be made from thick plate material and packing placed under the outer end to obtain the required height.

One important thing to do before using the faceplate at all is to check that it is true. Most are made of cast iron which is not the most stable of materials and it will frequently be found that a new faceplate in particular will be very slightly warped. To get it right is just a case of mounting it on the lathe and taking a very light skim across it, have the lathe running as slowly as possible when doing so. Being unstable there is every chance that after a period of time the faceplate will again warp and the operation will need to be repeated. It is most likely that this will be caused by the clamping nuts being tightened more in one area than others, thus applying extra

Another view of the flywheel, the outside rim is being machined, Note that packing has been put between the work and the faceplate so that the tool will not damage the latter.

This device is called a Keats Angle Plate, in use it is bolted to the faceplate and is extremely useful for holding odd shaped work.

pressure to that place.

Sometimes we get work to machine that it is difficult to set clamps on, if it has a good flat surface it can be stuck to the faceplate for machining. This can either be done with double-sided adhesive tape, but if so use a good strong type, like that used for holding carpet down, not the type used for sticking documents together. Alternatively an impact adhesive will work, if using the latter make sure that it is used in accordance with the maker's instructions, putting the two parts together too early can result in an inferior joint which could have disastrous results when machining starts.

Clamping work to a faceplate that is mounted on the lathe can be the dickens of a job. The work and clamps all want to slide down and as fast as one piece is tightened another falls down. For this reason it is a good idea to make up a little

device to fix the faceplate onto so that is can be laid flat when the work is bolted to it. All that is needed is a piece of bar with the end machined to simulate the nose of the lathe mandrel. This may mean cutting a large thread in some cases, an alternative being to machine the piece so that it is a good snug fit inside the faceplate thread. The device can be held in a vice and the work fitted at leisure, it is even possible to rotate the faceplate to make it easier to

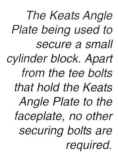

The Keats Angle Plate being used to secure a small cylinder block. Apart from the tee bolts that hold the Keats Angle Plate to the faceplate, no other securing bolts are required.

assemble things. If a small centre is made to fit the middle of the device, it will be possible to measure exactly the position of the work piece in the same way as if it was on the lathe.

Another useful idea is to fix the work to a piece of wood and in turn fit that to the faceplate. The best choice for this sort of work is a piece of stout plywood at least ½in or 12mm thick. The fact that it is made of layers of wood glued together with the grain going in opposite directions means that screws hold in it very well, although it is best to secure it to the faceplate with tee bolts and nuts rather than screw it. Frequently the bolts do get in the way when putting work in place and an advantage of using wood is that it is possible to be flexible. Perhaps using two bolts in places that will not interfere with anything, while also using wood screws in areas where the protrusion of a bolt would not be welcome. As a rule it is quite satisfactory to hold the work to the wood with stout wood screws, clamps being used in exactly the same way, as they would be if they were to be bolted in place.

Another advantage of the use of the wooden sub base is that if drilling or machining is to take place that penetrates right through the work, the tool runs into the wood instead of into the faceplate. Of course it is also possible in such cases to bolt the work to the faceplate and fit spacers underneath, that will also prevent the faceplate from being damaged.

Machining between centres

Another way of holding work is to support it between centres, it is a method to be recommended for obtaining extreme accuracy. A centre is inserted in small holes at either end of the work, one is held in the mandrel, the other in the tailstock. A gadget known as a carrier is put over the headstock end and clamped to the work. This is driven by a catch plate, which is rather like a small faceplate, without any slots and with a prong sticking out to locate with the extension on the lathe carrier. As the lathe rotates the peg on the catch plate pushes the carrier round, which in turn rotates the work. It is essential that the point where the tailstock centre enters the work is kept lubricated, otherwise there is not only a danger of it seizing up, but also the centre

*Below Left: A large nut converted into a lathe carrier by inserting a long bolt through one of the flats.**Below Right:** A more regular type of lathe carrier; designs vary according to manufacturer but all work on the same principle.*

Work held between centres possibly the most accurate way of machining.

will be ruined in no time at all. The centre must also be kept in contact as if it works loose at all the work will wobble and the finish will not be true. The centre at the tailstock end must be hard to prevent undue wear, some people prefer to use a rotating centre for tailstock support. Doing so saves any worries regarding lubrication but it may be found that when using that type of centre it is impossible to machine right to the end of the work. In certain circumstance that can also apply to an ordinary plain centre, it may not be possible to get the lathe tool in close enough to the work. If this is the case a half

If the work that is to be machined between centres is very small it is possible to make up a false driving plate and centre. The device is held in the ordinary three jaw chuck for machining and must not be moved as doing so will result in lack of accuracy.

centre can be used instead of the full one.

If one does not possess a catch plate the faceplate can be pressed into service, by simply putting a nut and bolt through one of the slots and allowing that to drive the dog. Carriers too can be improvised, a large nut for example, drilled centrally through one of the flats and a piece of round bar or a bolt inserted in the hole makes an excellent driving dog. Cap screws are particularly useful for this purpose, generally they are made of tougher steel than a normal bolt and in addition the round cap locates better with the catch plate than the hexagon head of a bolt.

The marking of the centres on the work piece can either be done with odd-leg callipers or alternatively the work can be set on vee blocks and either a surface gauge or scribing block used for marking. Ideally four marks should be made rotating the work ninety degrees between each, providing they are close to the centre, the true centre is easily found and can be marked with a centre punch. The centre punch mark is not really deep enough to support the work and it will need to be deepened with a centre drill.

Chapter 7

Tool holding

There are number of types of tool holder, or as they are more often referred to, tool posts, in general use. There are also many types that have been designed by individual lathe owners, wanting something that will suit their own particular requirements. These are far too numerous for it to be possible to describe all of them in this book and so only the more common types will be dealt with. As experience is gained in the use of a machine it may well be that an individual will have an idea for holding tools that will personally suit them. That is fine

Clamping Stud

Jack Screw

Thrust Washer

Packing

Top Slide

Basic single tool clamp

and as long as the device is capable of holding the type of tool in use as rigidly as is possible, there is no reason why it should not be so used.

Requirements
The basic requirement for any tool holding device is that it does just that, it holds the tool in place. It is essential that it can do so in such a manner that the tool is held rigidly and cannot move during operations. There is also a necessity for the height of the tool to be adjustable, one way or another and if possible that it can be set at different angles to the work. If in addition to this it is desirable that it will allow the tool to be set quickly as well as accurately.

Tool clamps
The most basic design is nothing more than a simple clamp that is tightened with a single screw and on some lathes this will be the only type of tool holder that is supplied. Height adjustment has to be made by inserting shims under the tool and to obtain the required setting angle it is simply a matter of placing the tool at the angle required. Usually tools of a larger cross section are used with this type of

Above: When tool posts are not adjustable for height, shims are packed underneath in order to get the tool in the correct place.
Right: A four way tool post, probably the most common of all, frequently this type of tool post is home made, providing a simple exercise in machining.

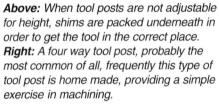

fixing; also it is common to use tool bits set in special holders. If several identical holders are available it is possible to grind the tools and set them to the required height and when another type of tool is required simply change the holder with the new tool already set to the required height. The clamp type tool holders have now largely fallen out of favour and yet the very fact

that a hefty tool or holder is needed gives it a rigidity that is not always there with some other methods. Possibly the main reason for its demise is that it is time consuming to set a tool in this way.

Single tool post

The single tool post is a block of metal, usually square. With a hole for the clamping bolt to pass through and usually a slot into which to fit the tool, which is held, firm with several bolts. At one time square headed bolts were always used for clamping the tool in place because the corners were quickly worn off hexagon heads, making it impossible to clamp the tool down tightly. Now it is much more usual to use cap head socket screws and to tighten them with a hexagon wrench. As a rule the height of

A tool post similar to the four way type except that a slot has been milled in one side to make it easier to take on and off the lathe.

the tool is adjusted by inserting shims. With some tool posts of this type instead of the tool being placed in a slot the block has a hole drilled in it and this accepts a tool made of round bar. The idea is particularly popular for use with boring tools. There is one disadvantage and that is the fact that it is essential that the tool be ground in such a way that when inserted in the hole, the cutting edge will be at the correct height in relation to the lathe mandrel.

The four way tool post

Similar to the tool post described above in as much as the tool post is basically a block of steel with a central hole for the bolt that holds it to the lathe. A tool post of this type is essentially square so that all the tools will be in the same position when it is rotated. This does not apply to the single tool post, which is best made in rectangular form. The four way post has slots on all four edges, instead of a single one thus in theory allowing four tools to be set and to be rotated so that whatever tool is required can be brought into use. As with the single post tools have to be adjusted for height by using shims. The idea can be very useful and save a great deal of time as it is no longer necessary to change tools for each operation, simply turn the post round to the required one. This is not always quite as successful as it sounds, in order for the idea to work it is necessary that the tool post be returned to the identical angle that each tool

requires. It is fine as long as the tool post can be set at exactly 90 deg to the mandrel as in that case all tools will return to the correct position. Should it be necessary for a particular tool to be set at a different angle other than 90 the only way to get it right is to make some sort of mark to identify the correct position. Some four way tool posts are equipped with an indexing device to overcome this problem.

One other problem that frequently occurs when using the four way tool post is that the tools in use overlap the ends of the slots and it is not possible to set a tool in the second one. The simple answer of course is to use shorter tools, but this too can have its problem as it may very well mean cutting high speed steel tools to the required length which is a very difficult task. With this in mind some tool posts are made with two slots, allowing a tool to be mounted on either side of the block and in some instances there is a slot on one side and a hole to accept round tools on the other.

A home made tool post that because of its bulk gives good support, while at the same time allowing height adjustment of the tool. This is achieved by the use of rotary cams that can be seen in the tool slot.

Height adjustable tool posts

When it comes to tool setting, particularly if a job requires a series of different tools, height adjustable tool posts are by far the easiest method of tool holding. There are two common types, one of these can be purchased the other will almost certainly need to be home made, let us deal with the latter type first.

They are based on the idea of moving the tool post up and down a central pillar that needs to be of sufficient diameter to allow the tool post to grip firmly on it. In other words it is not practical to fit the tool post on to an ordinary bolt of the type used to secure the normal tool post in position. This means the tool post consists of two parts, the post and the tool holder. The latter is split from the central hole to the outer

A height adjustable tool post that is interchangeable with others of the same type. The pillar is clamped tight to the top slide and the tool post mounted on that, it is tightened via the two screws at the rear and height is adjusted by the screws on the top which press on the cross slide to lift it as required.

The Gibraltar Tool Post designed by the late George Thomas. Its massive proportions and the fact that it is mounted on the cross slide rather than the top slide means that in use it is capable of being used both for heavy cutting and obtaining a fine finish. Photograph courtesy of Hemmingway Kits.

edge and the ensuing gap is capable of being closed by tightening a couple of screws. When it is tightened the central hole is reduced sufficiently in diameter to ensure the holder is clamped firmly to the post. There are one or two vertical screws through the holder that can be used to adjust the height when the clamping arrangement is loosened. These screws press on the cross slide, or top slide, depending on how the tool holder is mounted and in order to save damaging that any more than necessary it is usual to place a large diameter washer underneath. The tool holding arrangements follow the same pattern as with any other type of tool post.

As these are home made devices it is inevitable that there will be some individuality in their shape. The most common one is a shape that can only be described as semi-triangular and it although it is easily made from a mild steel block it is also possible to purchase castings for it.

The underside of a large and heavy tool post designed to fit at the rear of the cross slide. Because of its length and weight there would be a strain on the cross slide, this is prevented by adjustment of the studs that are mounted beneath the tool post.

The second version looks very much like the four tool type referred to above, except that it has the larger hole for the central pillar. Other than the shape it works on exactly the same principle as the previous version and because of its shape it is possible to mount two tools in it at the same time. It is not suitable for large lathes,

77

Tool posts that fit at the rear of the cross slide are useful for many jobs, although generally associated with parting off. The picture shows a small home made version with a short extension to allow extra room.

closing the gap round the pillar with the screws requires a considerable amount of pressure. This pressure is considerably reduced with the triangular form as there is less metal to be moved, nevertheless both types have their own particular advantages.

The type of adjustable tool holder that is usually purchased, although many people do make their own, works on a different principle altogether and consists of a block to which interchangeable tool holders are attached as required. The tool holders are made in such a way that they fit into dovetails in the main body, which is fitted with a means of tightening them to prevent movement. There are two methods of doing the clamping. The first is similar to the tool holders referred to above, closing a slot or slots in the body, by tightening a screw. Because of the pressure this requires the idea is fine on smaller lathes but on a large one the effort required to close the gap is considerable.

In the case of the second version of the tool post the tool holders are secured by means of a cam arrangement. A lever is used to rotate the cam and the arrangement is, as a result quicker to use than the previous version, as a single quick movement is all that is required. There is no great pressure needed for tightening either.

Elsewhere in this book readers have been told to ensure that as much vibration as possible is avoided and perhaps this is a disadvantage of using either of this type of adjustable height holder. No matter how tightly the holders are held there are additional areas that might allow some small movement to take place. They are therefore probably not as suitable for work involving heavy cutting as the one-piece

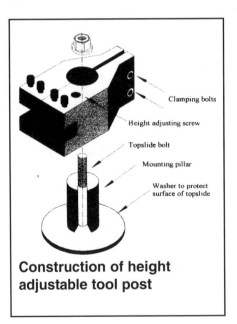

Clamping bolts

Height adjusting screw

Topslide bolt

Mounting pillar

Washer to protect surface of topslide

Construction of height adjustable tool post

type of holder. But their flexibility makes them ideal for lighter work.

Other ideas have been used for making it easy to adjust tool height and a simple arrangement that is quite effective is to fit the tool, which in this case must be made from round steel, into an off centre plug and to rotate it to the correct height. It is an idea that enables one to use a nice large tool and although a little awkward to adjust is very effective.

Dispense with the top slide

One of the main causes of a bad finish on work is the unwanted movement of the top slide. No matter how well it is adjusted it is difficult to obtain enough rigidity. The only way to be certain that there will be no movement or vibration is to tighten the gib strip as hard as it will go. That is all very well but it means that firstly it will be difficult or perhaps impossible to move and then it needs to be readjusted when there is next a need to use the slide. Many years ago, the late George Thomas realising that the top slide was a cause of problems designed a tool post that fitted straight on to the cross slide, thus avoiding the problem altogether. He called it the Gibraltar Tool Post and the design certainly had an appearance similar to the Rock of Gibraltar but more

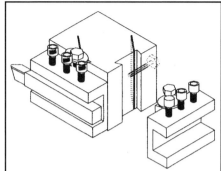

Height adjustable tool post

The tool holders fit between the vees on the main block and height is controlled by a bolt and washer on the tool holder. There is a split and bolts are used to clamp on the dovetails

importantly it works particularly well and work done when using it is far better than that done from the top slide. Castings are available for the original design, it is also possible to make something similar from any hefty chunk of metal and it is an exercise that is well worthwhile. It cannot of course be used for turning short tapers where the top slide is set to an angle but for normal turning operations it is ideal.

Chapter 8

Centres

The word centre can be very confusing to the newcomer working on a lathe, as it is one that is constantly reoccurring and has a number of meanings. For a start the lathe itself in called a centre lathe, even if it is generally shortened to just lathe.

We all of course know what the word itself means, it is the exact middle or centre of an object and it is used in this sense to describe the centre of the end of a round bar of metal but it also could mean an indentation that is not necessarily central. Confusing, isn't it? The following definitions may offer some assistance. Firstly as has already been established it can mean an indentation in a piece of metal, the indentation might be there for a number of reasons. The indentations might be made with a *centre* punch or maybe a *centre drill,* but they will be there as a means of locating

A Morse taper centre. These are available as soft or hard, the latter being used to support work from the tailstock and where a soft one would rapidly wear. Plenty of lubrication should always be applied when a centre is used for this purpose. Also shown is a half-centre, also used to support work from the tailstock but reduced in width to allow access for the cutting tool.

some particular point. Therefore it is possible for centre to mean any position on a piece of work that it is necessary to locate for a particular reason.

The name is also applied to pointed objects that fit either the headstock or tailstock of the lathe; there is a Morse taper or similar fitting at one end and a sixty-degree point at the other. They are almost invariably of steel and some are hardened. It is also possible to get half centres, in which case approximately half of the sixty-degree pointed end has been removed. The positions in which these centre fit are important, if fitted in the mandrel they are set at the true centre of rotation. In the tailstock they generally represent a matching position at the other end of the machine.

However it is usually possible to move the position of the tailstock in which case they become the end position of an angular line drawn from the true centre of the lathe. The usual reason for moving them in this way is to work on angular components. If the tailstock is moved we have the position where the line between its centre and that of the lathe forms one side of a triangle. As we know the second side is a straight line from the lathe centre, all that is needed is to know the length between what would have been the centre of the tailstock if it had not been moved and it is possible to calculate the angle of the line between lathe centre and the revised one in the tailstock.

Centre height refers to the exact centre of the lathe mandrel and is the measurement to which the cutting edges of tools are set.

Morse taper centres

Generally just referred to as centres, the morse taper centre is both a versatile and valuable accessory to the lathe. Accessory is really hardly the word for them as they are an absolute necessity and have a wide variety of uses. In their basic form there are two varieties, hard centre or soft centre. The hard type is used as a support for work and mostly held in the tailstock. It is hardened because the work is going to rotate on it and create a lot of wear. When it is being used for this purpose it must be kept well lubricated at the point of support. The soft centre is used to support work that is being driven and as it is to rotate with the work there is no need to harden it. As it is not hardened, it has a little bit of 'bite' into the work it is supporting. After a while the point on a centre becomes worn and it will be necessary to renew it, this can be done by simply machining it while it is fixed in the headstock.

The centres are pointed at one end, the other being a morse taper to fit into the headstock or tailstock of the lathe, No matter what size the morse taper, the taper of the point will be an included angle of sixty degrees. Traditionally the hard centre has been made from carbon steel and then hardened. More recently they have been produced in tungsten carbide which is a much more suitable material and far less prone to wear.

Half centres

When a morse taper is supporting work at the tailstock end it is very difficult to machine the end face of the work as the tool hits the circumference of the centre, before reaching the middle of the work, in those cases we use what are known as half centres. These can be bought but many people prefer to make their own, doing so is a useful way of using an old centre that has become well worn. Half centres have

A rotating centre that is also used to support work from the tailstock. However, because it rotates with the work, there is no need for lubrication of the actual centre.

the point reduced to nearly half the original diameter. Unfortunately it is not practical to reduce them to an absolute half as in doing so the support they supply would be lost. This means that the tool still cannot be taken absolutely to the centre of the work, but it will be very close and the diameter of the hole in which the half centre engages may be sufficient to allow the work to be completely faced off. How close one can get to the work centre will depend on the tool. A normally ground right hand turning tool will foul a half centre and so if facing off work that is supported in this way it will be necessary to grind a special tool to do the work.

Rotating centres

Centres set in ball races and mounted on morse tapers are known as rotating centres for the obvious reason that they rotate. They are used instead of hard centres to support work from the tailstock. There is less wear on a rotating centre than there is on a hard

centre. If they have an Achilles Heel it is the fact that the housing of the ball races is frequently too large to allow machining to take place at the very end of the bar. This apart they are an extremely useful accessory to have, unlike the ordinary hard centre, which can heat up to the point of seizing, this does not happen with the rotating centre.

Other centres

There are other types of centre used as supports, although they are not often seen. The most common is probably the pipe centre, which is a rotating centre but with a very large angled section, the angle of

Another type of rotating centre and known as a pipe centre as it is intended for supporting work with large bore.

The parts of a homemade pipe centre, showing the ball race that allows the head to rotate.

which is steeper than that of a normal centre. They are intended to support large diameter tubes or hollow work and are available in a variety of diameters. The reason for the steeper angle is to allow them to be brought nearer to the work than would be practical if the point was set to sixty degrees. It is essential when using one to make sure that the end of the work to be machined is running true, if it is uneven in any way the centre will throw it out of alignment and defeat the very purpose for which it was intended.

Virtually the reverse of a normal centre the hollow centre is used to support work that cannot be supported by a normally shaped centre. A typical example of this would be a piece of work that had a domed or pointed end. Sometimes a ball shaped centre might be used to support particular work, especially if the work has a large hole or is hollow and it is not unknown for a hollow centre to have a ball bearing temporarily set on the end in order to perform the same function.

Centre drills

If we try and start to drill a hole without some form of guidance, it is a million to one that the tip of the drill will wander away from where it was wanted and not only will we get a hole in the wrong place but in addition it will be neither truly round or accurate to size. We therefore start off by using a centre drill, the proper name is actually combination drill because it can be use for

The pipe centre in use. Note that, in this case, the pipe has previously been plugged with wood because the pipe is of thin, soft copper and applying pressure with the centre could possibly cause it to bell out.

Work supported at each end by a centre for machining.

countersinking as well as for starting holes. Centre drills are usually double ended with a parallel portion that has either a sixty or ninety degree angle at the end. It is rare to see the type with the ninety degree angle and most general tool stockists will only have those with a sixty degree angle. It is also possible to obtain centre drills that are described as bell type, the sixty degree cutting edged being slightly convex. The design is intended to make it easier to use for starting holes that are to be drilled. Whatever the angle or shape of the main section, there is a short parallel portion at the front and on that an angle of a hundred and eighteen degrees ending in a point. Both tapered sections are fluted and each tapered section has two cutting edges. Readers will no doubt have noted that the main angle of sixty degrees is the same as a standard countersink, the other the same as the point on a twist drill. Because the tool is sturdy and rigid it does not wander away from the starting location like a drill. Centre drills can also sometimes be used to correct holes, when a drill has moved off line and starts to wobble a little, they cannot deal with major misalignment however.

As well as being used to start drills, the short section of the centre drill is also the part that is used to make the centre necessary to support work while it is being machined, at the same time providing a grease filled cavity for lubricating the work while it is rotated. The angled section is a countersink and although it can be used to assist when starting a drill, it is not intended for deep penetration. Centre drills are available in a variety of diameters. both

metric and imperial. Sometimes it is necessary to start a hole in a position that the centre drill cannot reach. An example of this would be if the work has a recess and a hole is needed inside it. It is possible to purchase extra long centre drills to cope with such a situation. They would generally have to be ordered specially from an engineering tool stockist. As a good alternative in that case a piece of suitable sized rod can be drilled and a centre drill set in that to give the extra reach. The centre drill can either be secured in the hole with a grub screw, or held in place with a retaining compound such as Loctite.

The word centre again arises when we think of turning between centres, a subject that is dealt with elsewhere, but the

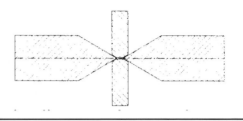

Re-lignment of centres after setting over the tailstock by lining the points within a countersunk hole made in both sides of a piece round bar.

operation involves finding the position on the work that is to be the centre of turning operations. It is customary to then centre punch the position and follow that with the use of a centre drill in order to obtain the recess in which to locate the lathe centre. The same expression will be found on drawings and relates to the distance between two specific points. It is known as the distance between centres and to the beginner all this can be very confusing. So there we are literally everything on a centre lathe revolves around various definition of the word centre.

Chapter 9

General machining operations

The most basic thing of all for successful turning operations of any sort is that the tool must be sharp, but in addition to that there are other important factors to be considered. Another golden rule for all machining that cannot be stressed enough is that all cutting operations should always be done as close to the chuck jaws as possible and there should also be as little overhang of the tool from the tool post as possible. Any overhang of either the work or the tool is going to lead to inaccuracy as well as a bad finish. Most readers will be aware of the principal of the lever, the longer the end to which power is applied is away from the object to be moved, the easier it will be to move it. The same applies to turning operations, as soon as a tool is

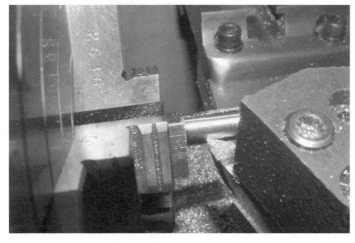

Machining operations should always be carried out as close to the chuck as is feasible and with as little overhang of the tool as is possible.

An easily-made gauge for setting a tool to the correct height.

pushed into the work the movement of that work will be away from the tool. The further away from the chuck, or other means of work holding, the greater leverage that is applied the more that movement is likely to occur. In the case of the cutting tool, there is a tendency for the tool to be forced down and again if the cutting edge is some distance from the secure point, which in this case is the tool post, the greater the tendency of the tool to move down.

The tool must be set at the correct height and it is essential that it be presented to the work at a working angle, this should relate to the angles to which the tool has been ground. For example, if we set an ordinary left hand tool at an angle, instead of at ninety degrees to the work, the cutting angles then change as they are directly related to the work piece. It may well be that by doing this the end rake becomes non-existent and may even rub on the work as the cut is being taken. One of the advantages of using high speed tools which one grinds for oneself is that if for any reason it is desirable to set the tool at a

The sequence of photographs shows how centre height can be obtained by using a 6-in (15-cm) steel rule and simply laying it on the metal bar and winding the tool to touch it. On the left is what happens if the toll is set too high; the central photograph shows the toll set too low; the tool set to the correct height is at the right.

A three-point steady. All three-point steadies are bolted to the lathe bed and should be adjusted to provide support as near to where machining will take place as is possible.

different angle, other faces can be ground to suit the new set up.

It is not of course always possible to adhere to this principle, for example it may be that a long length of metal is to be machined, in which case it must be supported. This support may take the form of a centre in the extreme end, or the use of a steady, depending on the length of the material. Whichever it is support is essential. There may well also be instances when it is not possible to use the cutting edge of the tool as close to the tool post as we would like, because of some obstruction or other. Sometimes it may be possible to support the tool in some way, if not then the tool used should be as hefty as possible. That way at least the actual tool is less likely to flex than a smaller one, although the use of a hefty tool will not entirely stop movement being transferred to the tool post and cross slide.

Steadies

Too much emphasis cannot be placed on the need to have good support for the work and when it is possible and practical to do so it should be supported by a steady to avoid the tendency of the work to move away from the tool. Anything that supports the work can be considered as a steady and all sorts of items have and are used to do the job. We generally use the term to describe two types of devices. The three-point steady bolts to the bed of the lathe and has three arms that are adjusted so

that each offers support to the work. The arms must be made of a material that is softer than the work that it is being used to support; otherwise it can cause scratches or even deep scores on the job. There is inevitably some amount of friction involved when using the three-point steady and so some form of lubrication should be applied. Some people have solved the problem by fitting small diameter free running wheels at the end of the arms. The idea serves two purposes, less lubrication is needed and there is much less chance of damage to the work.

The two-point steady works in a different way as it is bolted to the saddle and moves along with it. This means that the steady is close to the point of contact between the cutting tool and the work. It largely will depend on the type of work that

89

A two point steady of a type that is bolted to the saddle and moves with it, thus providing support as close to the tool as possible. When machining long work, consider the use of both types of steady in order to get the best results.

Both types of steady are readily available from any good tool stockist, for those who enjoy making as much equipment as possible it is possible to buy suitable castings. They are also very easy to fabricate, there is no need for a cast body, steadies can also easily be fabricated from mild steel sheet

The size of steady will depend on the size of lathe, but even when using a small machine, turning down very small diameter work calls for a different approach. Apart from anything else, very few steadies will be capable of supporting very small diameter bar stock. In this case it will be a case of the operator having to provide his or her own device and as a general rule it will take the form of some sort of soft collar is being carried out as to which type of steady should be used, but they should be used whenever it is possible to do so. It is just as essential to lubricate the pressure points from a two point steady as it is when using the three point one.

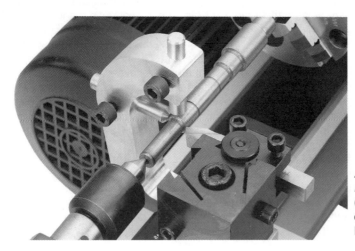

A two-point steady in use. (Photograph courtesy of Proxxon Tools.)

in which the material will run. In the interests of accuracy the device, or at least the supporting material should be made in situ, it can then reasonably be assumed to be accurately lined up with the lathe centres.

Two other things are also particularly important, firstly as has already been pointed out the cutting tip of the tool must be set at the correct height and if this is not done not only will it be difficult to do the machining but also the end result will be far from good. The second important thing is that the machine should run at the correct speed for the material that is being machined. In practice unless one has a machine with infinitely variable speeds and a suitable tachometer device, obtaining the correct speed is not a practical proposition. Most lathes have a range of speeds and we are therefore governed by that range, so although every effort must be made to get as near to the correct speed as possible, this is probably the best that can be done.

Anyone who did an apprenticeship in engineering some years back would have had the formula with which to obtain the correct speed drummed into them many times. It is $\dfrac{CSx12}{\Pi D}$ (Cs = cutting speed) (D = diameter): for metric work $\dfrac{CSx1000}{\Pi D}$ There are for example many types of steel and this does not mean just mild steel, carbon steel, etc. One only has to look through a book of steel specifications to realise that there are hundreds of different types and in theory at least, each has its own cutting speed. We cannot of course possibly know the cutting speed of every one and indeed more often than not the home machinist will have no idea of the type of material in use. He or she will just buy a length of mild steel and is highly unlikely to ask the stockist what the cutting speed is, some would not know the answer anyway. Model engineers are usually only too happy to accept the gift of a piece of metal from a fellow club member and most definitely will not be quizzing him or her about the cutting speed. The only way therefore when calculating the speed to set the lathe is to take an average figure for various metals and use that. Tables have been prepared to assist readers to find a suitable speed and they do just that, they give an average for particular types of metal.

If we work as near as we are able to the recognised figures for setting lathe speeds for cutting then it follows that as the diameter of the work decreases, the rotational speed should increase. While it is impractical to constantly make changes it is advisable when a considerable decrease in diameter is being machined, to stop a couple of times and increase the speed to something more suitable to the revised diameter. There are also additional factors that need to be taken into account; one of these is the depth of cut, as a deeper cut requires a slower speed than a light one. A very fine finishing cut is always required to enable a good smooth finish to the work and to get a good finish; the lathe should be run at a faster speed than it is for operations to reduce the diameter. In spite of all these variations it is useful to have some idea of a figure to work on, rather than relying entirely on ones own judgement

All of this sounds most confusing to the beginner to machining and therefore it is best to use the published figures as a guide

Nomogram providing a rough guide for calculating the required speed when machining various metals. Using a straight edge, place one end on the name of the metal and the other on its diameter. The figure in the middle column that the edge crosses is the required speed. Where there is no direct alignment, use the next lowest figure. This guide is intended for high-speed steel tools – so double the speed if using carbide tipped versions.

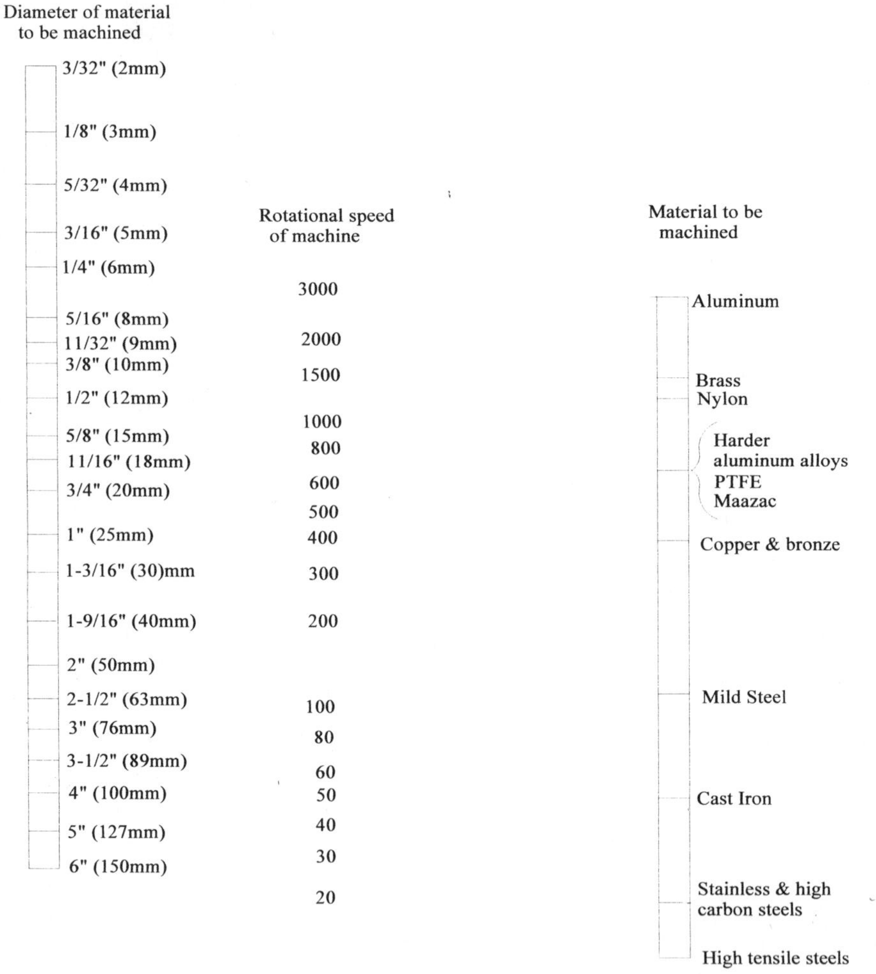

Diameter of material
to be machined

3/32" (2mm)

1/8" (3mm)

5/32" (4mm)

Diameter of material to be machined	Rotational speed of machine	Material to be machined
3/16" (5mm)		
1/4" (6mm)		
	3000	Aluminum
5/16" (8mm)		
11/32" (9mm)	2000	
3/8" (10mm)	1500	
1/2" (12mm)		Brass / Nylon
	1000	
5/8" (15mm)	800	Harder aluminum alloys
11/16" (18mm)		
3/4" (20mm)	600	PTFE
	500	Maazac
1" (25mm)	400	Copper & bronze
1-3/16" (30)mm	300	
1-9/16" (40mm)	200	
2" (50mm)		
2-1/2" (63mm)	100	Mild Steel
3" (76mm)	80	
3-1/2" (89mm)	60	
4" (100mm)	50	Cast Iron
5" (127mm)	40	
6" (150mm)	30	
	20	Stainless & high carbon steels
		High tensile steels

A three-point steady in use.

and increase or decrease the lathe speed a little as required. Listening to the work is probably the best guide of all, if things sound laboured or there is a screeching or chattering, then something is wrong and it is worthwhile trying a different speed. It should be kept in mind that such noises may have causes other than spindle speeds. A badly set tool, or badly adjusted slides frequently cause chatter; it can also be caused by too much tool or work overhang, or lack of lubrication, a subject that we will come to shortly. So don't rely just on a change of speed to cure matters.

Those who have to rely on a permutation of belt positions and therefore have to accept the nearest available mandrel speed, will find it is best to work at a slightly lower rate rather than a higher one. Running the lathe too fast is likely to result in overheating the cutting edge of the tool and will then rapidly become blunt, this particularly applies when using carbon or high-speed tools. It may even be impossible to resharpen them as they may well have lost their hardness as a result of the heat.

Tool height

If a tool is set too high it will wear the metal away rather than cut it, resulting in a bad finish and rapid tool wear as well as overheating of both tool and work. If it is set too low there will be a similar effect but in this case the top of the tool will be rubbing on the work, the result again will be to cause wear and overheating. If we accept the advice given in the technical books, for ordinary straight cutting along a bar the tool should be set at five degrees above centre height, this equates to 1mm per 20mm or 3/64 per inch of diameter. For boring, taper turning and screw cutting it must be set exactly on centre height. However setting a tool at five degree above centre height each time is not that easy, it requires careful measurement and so for general work the advice is to set the tool exactly on centre height. There are numerous ways of setting the tool at the correct height, it can be lined to the point of a centre, either in the

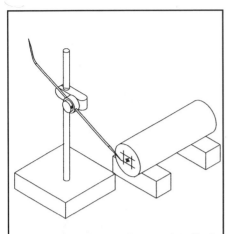

A surface gauge can be used to find the the centre of a bar for turning between centres.

headstock or the tailstock, in which case the lining up should be viewed from level with the centre, not from above as that can give a distorted view. Many people like to make a gauge, which need be nothing more than a pillar in a base, with a short length of bar on it. The bar is machined flat at the required height and when the tool is brought to it whether or not it is level can quickly be established by running a finger over it. A quick method that gives a good degree of accuracy when round bar is to be machined is to place a length of thin steel against the circumference of the metal and gently squeeze it against it with the tool. If the cutting edge is at centre height the strip will remain at ninety degrees, if not it will tip at an angle. A six-inch (150mm) rule is commonly pressed into service for this job.

Setting up

If the work piece is a round bar then generally there is no need to set up. The bar is simply put in the chuck, tightened and away you go. If a step is to be machined in the bar nothing else may be needed as once the first cut is taken a measurement can be taken from it and worked to until the correct diameter is obtained. The cross slide dial should be used to establish the amount of metal that is being removed, rather than constantly taking measurements from the metal itself. One important thing to remember is that every movement of the cross slide removes twice the amount of metal than is indicated by the dial, it is something that is all too easily forgotten and has all too frequently been the reason for a job being spoiled.

If the work happens to be on a piece of rectangular stock or perhaps a casting then it will be necessary to set it up to run true. What is needed is to ensure that the centre of the section to be machined is absolutely true to what will become the diameter, or if a hole is to be drilled or bored the centre of what will be the hole has to be accurately located.

The first thing is to physically mark the required position and it is usually necessary to do this away from the machine and having found the position to centre punch it. The author actually goes a little further than centre punching, using an Archimedean drill to improve the centre punch mark. The reason for this is because when a centre punch is used not only does it make an indent in the work but it also raises the edges of the metal making them uneven and it is possible for the edge to cause some inaccuracy. The rough edge can be filed off but this usually means that only a shallow indentation is left. An Archimedean drill for anyone that has never seen one is a chuck that is rotated by

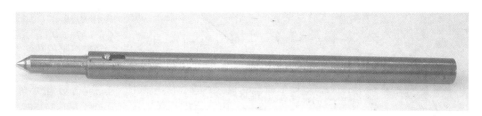

A home made centre finder or wobbler the point has been hardened to prevent undue wear.

moving a nut up and down a thread. It is frequently used with a spade type of drill and at one time the tool was very popular with woodworkers but with modern equipment available has now gone out of fashion. They are however still readily availably at hobby suppliers and are worth the small investment involved.

Assuming the necessary indentation has been made, whether by punch or drill, the point of a device called a centre finder is located in the indentation, the other end being supported in the tailstock. Centre finders are more likely to be found under the names of wigglers or wobblers. Some centre finders can be put in the tailstock chuck, others must rest in a hard centre. The work is then rotated by hand and the movement of the probe of the centre finder observed and adjustments made to the chuck jaws moving the work until the point of the probe is running perfectly true.

It is possible to see how accurately the probe is running by bringing the point of a scriber in a surface gauge to it. For greater accuracy it should be checked with a clock gauge which is as the name suggests, a clock. It has a probe that is brought into contact with the wiggler and any movement is registered on the clock. Using a clock gauge makes it possible to set work up very accurately indeed. Pictures relating to setting up in this fashion can be found in Chapter 5.

Frequently it will be necessary to know the position of the tool in relation to the work, trying to view whether or not the cutting edge is in contact with the surface, using either the naked eye or even with a magnifying glass is extremely difficult, it is wise therefore to use some sort of guide. Placing a very thin feeler gauge between the cutting edge and the work is quite a good way, but there may be some slight discrepancy in the final outcome. It is possible for the metal of the feeler gauge to compress a little when put under pressure by the contact of the tool, so only minimal pressure should be applied. Feeler gauges were things that at one time would

A commercially made centre finder. These are usually sold as sets with interchangeable probes.

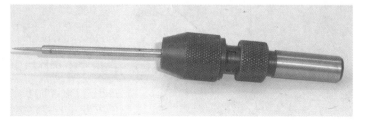

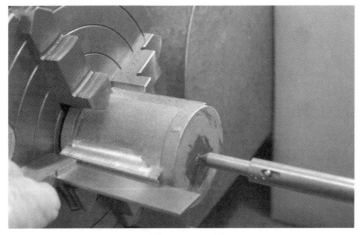

Setting up a casting using a 'wobbler'. In this case, because of the awkward shape and lack of a truly flat surface, a square is used against the face of the chuck to get the correct alignment.

be found in most home workshops but this is now less likely as home car maintenance becomes more difficult. If they are not available, cut a small piece from a food can, measure the thickness and use that, it is just as good as the feeler gauge.

A little trick that was used by old time turners gives even greater accuracy than the feeler gauge idea. It is to put a small piece of tissue paper about ½in or 12mm square on the work and put a touch of oil on it to ensure it sticks to the surface and is absolutely in contact with it and not wrinkled in any way. Start the machine and wind in the tool, when it picks up the paper the tool will be in contact with the surface of the work and needs no further adjustment. The paper that was traditionally used was a cigarette paper as used for rolling ones own cigarettes, it being exactly the thickness required.

Cutting fluids

We use cutting fluids when machining to keep the work and tool cool, thus preventing the tool from heating up to the point where

Setting up using a 'wobbler' and surface gauge. A piece of white paper has been laid on the lathe bed to make it easier to identify any off-centre movement.

96

its hardness might be lost. They also provide lubrication to the cutting edge of the tool, preventing a build up of material on the edge. This also reduces wear on the cutting edge of the tool, so that it will last longer without the need for sharpening. The cutting fluid should not be in any way corrosive so that at a later stage it will cause damage to the work or machine. Different materials require a different approach and for a start under no circumstances should a lubricant ever be used when machining cast iron. Doing so will cause the iron to harden and make it difficult if not impossible to machine. The material contains its own lubricant in the form of graphite and therefore needs nothing extra.

For general machining work in industry soluble oil, frequently known as suds, is used, as the name suggests it is an oil that is mixed with water to make a solution, the strength of which will depend on the grade and manufacture of the oil. It is possible to mix it with water as it is a vegetable oil, rather than a mineral one such as we use in our cars and which would only separate if attempts were made to mix water with it. Unfortunately being a vegetable oil it does tend to smell rather, in spite of the manufacturers' attempt to mix in other solutions that disguise that smell. In addition if left for long periods of time it can actually go bad, in which case the smell becomes unbearable. Nevertheless it is possibly the best form of coolant, particularly when used on steel and in industry it is applied via a pumped system

that keeps a steady flow of liquid continually on the work. The oil is taken from a sump under the machine, and falls into a tray from whence it flows through a drain to be returned to the sump. Pumping it means that there is not always a great deal of control over the flow, with the result that it tends to splash all over the place, usually including splashing the operator, he or she then goes away not exactly smelling like roses.

It is possible to purchase a pumped system for most makes of lathe if one so wishes and the same system of taking the oil from a sump and returning it via a drain is used. The alternative if one wishes to use soluble oil is to make up some form of drip feed that will allow a steady supply of liquid to fall on the work, the end result is almost as good as when a pump is used. The used oil will still need to be drained into a container and then returned by the operator to the original container on the lathe. It is usual to make ones own system for this method and as a rule consists of nothing more than a container on a stand with a tube that will go above the work, a simple tap provides the control of the amount of

Setting up a rectangular casting. In this case a 'wobbler' is not required as the casting is adjusted until each edge just touches the pointer.

A clock gauge being used to set up a casting. The stand for the clock is magnetic and is placed on the lathe bed where it remains secure.

fluid being dispensed. All fluid being returned in this way should be well strained before use in order to remove any stray pieces of swarf that may have worked its way in.

Other methods of dispensing soluble oil are frequently used, these include, filling a spray container of the type used for household polishes and cleaning materials, this is then sprayed on the work as required. It is not usually practical to recover the used oil in this case, but as considerably less is used than with the previously mentioned methods it possibly doesn't matter too much. The full amount of used oil is never completely recovered anyway. Some

Setting up does not only mean finding the centre of work, in this photograph the surface gauge is being used to check that the edge of the flywheel is consistently at ninety degrees to the axis of the lathe.

98

Machining across the end of the work is known, as facing, only light cuts should be used for this operation.

people also like to use a brush, dipping it in a tin of the liquid and then applying the brush directly to the work. While this method is quite popular the hairs of the brush tend to become entangled between the tool and the work. The result is a gradually denuded brush and a lot of brush hairs all over the lathe.

In the home workshop it is rare to find work being carried out with the same intensity as it is in industry and in order to prevent too much mess, alternatives to soluble oil are frequently used. One alternative is neat cutting oil, usually applied by hand with a drip feed from a container. These oils work on the principle of cooling

Another example of facing, this time on a casting to leave a boss.

99

An example of the use of the jaws of a four jaw chuck used in reverse to accommodate a large diameter flywheel.

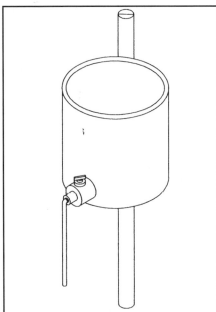

A simple arrangement of a tin can, a tap and a piece of tubing can be used to drip-feed coolant

the work by a rapid evaporation, somewhat on the lines of a refrigerator. The oil can be purchased and usually comes in a bottle suitable for allowing it to be used directly. Larger quantities can be purchased in which case it has to be transferred to a suitable container in order to use it. There are numerous versions of this type of coolant, some of which appear to be better than others, so selection of the type to use must be a personal choice. As the cooling effect is gained by evaporation of the oil they can give off fumes, these may or may not be toxic, but in many instances they are unpleasant when breathed in by the operator and it might be sensible to use a facemask when using it.

It is possible to make ones own neat cutting oil, it requires two parts of machine oil (old fashioned SAE 30 is ideal) to one part of white spirit, and this is put in a container and shaken vigorously until thoroughly mixed. About an eggcup full of washing up liquid is then added and the mixture is ready to be applied by whatever means suits the operator.

Both soluble oil and some neat cutting oils can create rust, so it is advisable to clean the machine after their use.

With the use of tipped tools there is less, if any need for cutting fluids to be used. The tools themselves will stand a considerably higher temperature before they come to any harm than high speed steel will. However it must be borne in mind that it is not only the tool that heats up but the work also, so any decision whether or not to use a cutting oil because a tipped tool is in use must also have in mind the effect it will have on the work piece.

Work should be carefully planned and it should be done in such a way that the larger diameters are machined first, leaving smaller ones until the last. Most lathes have a self-acting mechanism with which to traverse the work and for normal operations it is best to set this to as fine a feed as possible. Not everyone likes to use it and there are some highly skilled machinists that believe they can get a better result when traversing the saddle by hand. Only experience will enable one to decide which way of doing things is most satisfactory. Whether the self-act is used or not, all operations involving taking a parallel cut should be done by operating the saddle, never use the top slide, except if making a taper.

It is not possible to tell readers how deep the cut should be, there are too many variations, such as type of material, suitability of the machine, etc. The only thing that can be said is that more often than not taking an excessively deep cut, is self-defeating, it will be quicker to take two of half the proposed depth than one deep one. The speed at which the tool can travel is to some extent dictated by the depth of the cut and this should be borne in mind. In addition a very deep cut is likely to cause a certain amount of flexing on the tool resulting in lack of accuracy.

Having machined the work to within three millimetres or 1/8in final cuts of no more than initially 1mm or 1/64in should be taken followed by 0.25mm or 0.001in for the final two cuts. There is quite a controversy as to whether it is best to take final cuts at a high speed or a very slow one. Most textbooks suggest that the speed of the lathe should be increased by about 50% for the final cuts. Many people disagree with this and prefer to finish at a very slow speed and like most matters when it comes to turning operations it is a matter of finding what suits oneself and sticking to it.

The beginner when working with a lathe often finds it difficult to get both a good finish and accuracy and end up resorting to filing the last stages and then possibly using an abrasive cloth or paper for the final finish. Doing this is very bad practice as it will be almost impossible to file the work perfectly evenly and the result will be a piece that does not measure consistently throughout its length. Neither file or abrasive cloth is likely to give a really good final result it is far better to use a good sharp tool and a very fine cut which will give both accuracy and finish.

Facing

Facing is the name given to the operation of running the tool across the work rather than along it and the same parameters apply to doing that as apply to normal turning operations. The tool must be sharp, at the correct height and the lathe running

With a little thought a lathe can be used to accurately machine shapes of all sorts. Here we see the two legs of a casting being machined so that both are identical in length.

at the right speed. Facing cuts should always be on the fine side as a great deal of pressure is being applied and it means that the work will always have a tendency to try and move away from the tool. The deeper the cut the more that pressure is applied and the end result will be a poor finish.

Interrupted cuts

An interrupted cut is one where the tool will not be in contact with the work during the whole time. A typical instance might be when deliberately machining work off centre, such as if making a cam. It is only common sense to ensure that when the tool does make contact with the work it does so very lightly and continues to do so as long as the cut is interrupted. Carbide is particularly prone to breaking under these circumstances and it is better to use high-speed steel tools for this sort of work. There are occasions when work can involve a lot of this sort of work, such as if machining square or hexagon bar and particularly when machining castings,

Machining cast iron

In theory when machining cast iron it is desirable to remove the top surface in one cut. That surface is invariably much harder that the inside and if it is possible to cut through it in one go so well and good. If a round bar of centrifugally cast iron is being machined it may be possible to do so, otherwise it is highly unlikely that it can be done. This means of course that the tool may be subject to alternate hard and soft surfaces and the lathe speed will need to be adjusted to deal with the hard surface rather than the softer one. Sometimes when machining cast iron a hard spot will be encountered and it is a problem to know what can be done about it. Hard spots are caused by a section of the casting cooling too rapidly when it is removed from the mould. Sometimes by using a tipped tool it is possible to machine through the hard spot, frequently it is not possible to do so and unfortunately the work will have to be abandoned. Foundries will always replace such castings but it is annoying, particularly as invariably the hard spot only seems to

Damage to a tool with cemented tip, caused by using it for interrupted cuts.

show itself after most of the work has been completed.

Machining small diameters

How small is small? Our conception of small appears to be relative, to some people a screw of 3mm or 1/8in diameter is very small, and to others it can be comparatively large. To some extent the view taken depends on the type of work in which one is interested. The screw referred to above would be enormous to a watchmaker and he or she would be most unlikely to ever use such a thing. To someone restoring vintage vehicles, such as a traction engine that size of screw would be thought to be extremely fiddly.

Machining, small diameters can be very difficult and must always be considered in relation to their length. If we stick with the 3mm diameter then a length of 6 or even 12mm would not be too much of a problem, make the length 75mm or 3in and it is a different story altogether. The pressure applied by the cutting tool at the

extreme end will prove too much with the result that the work will bend and become impossible to machine. It is not impossible to do but to be successful requires certain precautions to be taken.

The first essential is a very sharp tool and it is desirable that the cutting angle is increased in order that there is plenty of clearance. Some form of steady is also necessary in order to be successful, the usual type is unlikely to prove to be suitable and so we must think in terms of manufacturing something suitable. Generally this will take the form of a support that is situated as close as possible to the

Machining between centres is the most accurate method of working.

A steady for maching small diameters. It bolts to the carriage and the hole has a piece of material inserted. The material depends on what is to be machined, if it were steel the insert could be brass, if brass was to be machined it would be PTFE or nylon. The insert is then drilled the same diameter as the work to be machined, using a drill in the lathe chuck.

cutting tool. Suitable devices are shown in the accompanying illustrations and it will be found that a type will suit one person while somebody else will prefer something entirely different No matter how good the support is it is necessary to proceed with the lightest of cuts and with the lathe rotating at quite high speeds. It iş also necessary to ensure that there is good swarf clearance. Using a suitable steadying device plus correctly prepared tools there is no reason why the machining of very small diameter should be a problem.

Frequently we are likely to need to make several items having small diameters, they could be screws, pins or a variety of other things. This often involves constant setting and resetting of the steadying device, something that is both time consuming and boring. Therefore if a number of items are required it is well worthwhile making special tooling to do the job and while it may sound time consuming to make tools in the long run it can be well worthwhile and in addition they can always be saved for use at a later date.

Running down tools, sometimes referred to, as rose bits are an easy way of machining very small diameters.

A running down bit in use. Because it encompasses the whole diameter of the work, providing there is no great overhang from the chuck a steady is not required.

Possibly the easiest and yet most reliable device is known as a running down tool, sometimes called a rose bit, because of the shape of the cutting edges, It consists of a short length of silver steel, with a hole drilled right through, the diameter of which is the one that it is intended to machine to. Cutting edges are filed in one end and one or two relief holes to allow the swarf to escape are made on the outer edge. The tool is hardened and tempered and then used from the tailstock chuck, being run slowly over the rotating work piece. The cutting edges reduce it to the size of the holes. It is particularly useful for making screws, as the hole that is drilled can be the same size as the outside diameter of the screw.

Using mandrels

Machining the outside of a piece that has a hole in the centre is best done on a mandrel. There are several types of mandrel and usually they will have to be home made to fit the job. Simply take a piece of metal of a larger diameter than the hole through the work, put it in the chuck and machine a spigot that is a close fit in the hole in the work. There will be two ways to secure the work to the spigot, or mandrel as it has now become, Either a short additional length can be machined and threaded to accept a nut than can then be tightened to hold the work secure. The additional length for the bolt must be machined at the same time as the making of the mandrel. Under no circumstances should the mandrel be removed from the chuck until it has been used for the job for which it was intended, it will be to say the least very difficult to get it true again once it has been removed. An alternative to the bolt is to drill the mandrel and tap it to accept a screw or bolt that also will tighten up and hold the work secure.

If the diameter of the work is large in comparison to the hole in it support should be given at the back. This means that either the mandrel should be made from a piece of large diameter stock or a thick washer of a suitable diameter should be put between it and the work. A typical example of this would be if machining a model

105

locomotive driving wheel or perhaps a flywheel for a stationary engine or traction engine model. At the outer edges there will be pressure for the work to move away from the tool resulting in a wobbly wheel, this can be prevented if thought is given to the support required from the mandrel.

Using round stock to make a mandrel to machine a few items may be the best way of doing the job, but some people might think it a waste of material as much of the stock is wasted. If a length of hexagon bar is used instead of round and a mark is made on the flat adjacent to the number one jaw, it should be possible to use the mandrel time and time again by ensuring it is replaced in exactly the same position by reference to the mark and jaw number. There is one small proviso to this, the idea is fine and works very well but to be successful it is essential that the chuck is perfectly clean before making the mandrel and each time it is used as well. A small piece of stray swarf or dirt will be sufficient to stop the jaws opening and closing accurately, so a thorough clean out is essential.

Chapter 10

Parting off

Always considered one of the trickiest of operations on a lathe and to some people a positively frightening prospect, parting off calls for particular care. Providing the job is dealt with sensibly there is absolutely nothing at all to be afraid of and it is possible to part off thick steel bar, without having any worries. Doing so requires that the job be thought out before commencing and that it is approached with care and common sense.

It will be seen that the blades of the tools used are made from a very thin section of material, which makes them particularly vulnerable. Tools made from high-speed steel now generally are tapered from top to bottom, with a front rake of about ten degrees. Mostly they have no top rake at all, although at one time this was considered essential, some people still prefer to grind this clearance rake on the tool as in their opinion it does offer assistance in swarf clearance. Whether to adopt the idea or not must be a personal choice. One thing against it and probably a reason why it is no longer carried out is that once the tool has become blunt if it is ground, the cutting tip is no longer at centre height. There is no reason why the height of the tool cannot be corrected after

A common type of parting tool, the blade extending along the tool holder and being secured with a nut. The complete holder is designed for mounting in a tool post.

The reverse side of the parting tool holder, showing the nut and stud for holding the blade firm during operations.

sharpening, although there may be a limit as to how many times that can be done.

Blades that are sold commercially are intended to be fitted in holders, some of which are designed to accept only blades supplied by that manufacturer and this could make the blades non-interchangeable. The difference in design are not great, some blades have a taper along the top edge and the matching holders are designed with a lip to accept it. Others are flat at the top but have an angle at the bottom that clips into a groove in the holder and the blade is held in place by a flat plate fitted on the holder and designed to push the blade into the groove. This does not apply to all holders and blades, many of which are universal fitting. Holders may also vary in other ways and some having a side bar for holding in the tool post; others are designed for the whole body to be held there.

At one time all parting off blades were considerably wider than many are today, they were mainly made of high carbon steel rather than high speed which meant that the cutting edge would wear away much more quickly and a thin blade would need constant sharpening. As the cutting edge was much wider there would be a clearance rake from the front to the rear of the blade as well as a rake from the top edge to the bottom, on each side of it. That top to bottom rake is still found on modern blades The other one has largely disappeared and it might be due to the fact that the wider blade meant a heavier chip was being removed, something that applies less when the blade is narrower.

Blades are available in a number of depths and width is usually dependent on the depth. The deeper the cross section of the blade the thinner it is likely to be and

A different and slightly unusual parting off tool and holder, in this case the blade is designed to pass through a slot.

A parting off tool with a cemented tip, the shape of the tool is identical to those made from high speed steel, without the benefit of the carbide tip.

obviously this also applies to their holders. If heavy work is to be carried out it is best to use a thicker and deeper blade, as it gives greater strength. The wider blade obviously makes a wider cut and it will remove more metal; metal which is just waste. This makes it tempting to use the narrowest blade possible and it is a nice thought that by doing so one is saving valuable material. However, this is not necessarily an economy measure as the narrower the blade the easier it will break and the quicker it will become blunt. Also the wider and deeper the blade, the greater its bulk and the more it will absorb heat, and heating up of the blade is likely to result in it jamming in the work and possibly breaking as well. Even so as one becomes more proficient in the art of parting off so it becomes easier to use narrower blades.

As with all cutting tools it is essential to keep parting off tools sharp. This is particularly so because of the nature of the cut that is being applied, where the whole width of the cutting edge is in continuous use, sharpness is even more important. Honing is also essential as ragged bits left on the cutting edge can spoil the finish of the cut. Very often the cut made when parting off is going to be part of the finished job so it all requires a great deal of care. As purchased some blades do not have the cutting edge ground the length of steel having been simply guillotined off to size. In the case of small blades it is easy to miss seeing this and to try and use the blade as bought. All new blades should be ground before use, as the finish on a new blade is rarely good enough to enable clean accurate work to take place.

In addition to high-speed steel tools as described above they are available made

A parting off tool with replaceable tip that is held in place with a clamping arrangement.

109

Ultra close up view showing how the sides of a parting off tool are tapered to give clearance.

from tungsten carbide and cobalt as well as with cemented carbide tips. It will be found with some cemented tip tools that the tip actually overhangs the body of the tool meaning that where the tip meets the main body, the main part is narrower. And means that there is less chance of the tool jamming in use. Parting tools with throwaway tips are also available and have the advantage that the tip is correctly ground and in most cases ground with a chip breaker. Chip breakers are discussed in the chapter on tools, they break up the swarf, and a good

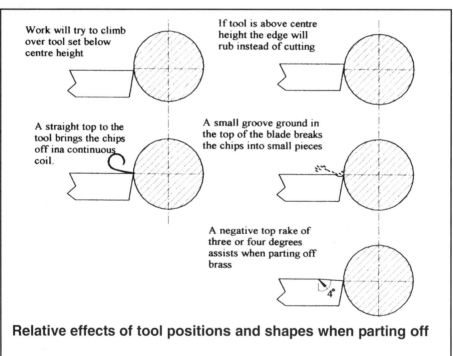

Work will try to climb over tool set below centre height

If tool is above centre height the edge will rub instead of cutting

A straight top to the tool brings the chips off in a continuous coil.

A small groove ground in the top of the blade breaks the chips into small pieces

A negative top rake of three or four degrees assists when parting off brass

4°

Relative effects of tool positions and shapes when parting off

Parting off is best carried out from the rear of the lathe as this puts the pressure down towards the bed, thus giving less vibration. The picture shows a tool holder specially designed for such work. Note the heavy construction of both tool holder and cross slide extension and the fact that the parting off blade must be mounted upside down for this type of operation. Although mounted upside down it is still essential that the blade is at the correct height and at ninety degrees to the work.

thing when parting off as a build up of a single chip coming from a parting tool is a recipe for disaster.

Before starting to part off work give some thought to the job, decide how deep a cut is required and what the material that is being worked on is. A different approach has to be made according to the metal being cut as well as the diameter. Generally speaking it will be easier to cut brass than steel, but aluminium another soft metal can be difficult to deal with.

There are two main difficulties in any parting off operation, one of the biggest is the build up of heat, the result of which can be the expansion of the work piece, causing the groove that has already been made to clamp on to the tool and seize up. The usual effect is to simply bring the lathe to a stop but it is also possible for the blade of the parting tool to snap off. Even if one is lucky and the tool remains in one piece there is the difficulty of removing it from the groove

and this is not always easy. Often combined with the heat expansion there is a swarf build up that locks the tool solid in the work, it might be possible to pull the lathe in reverse by hand and relieve matters in that way. Sometimes using a rocking motion will do the trick, but it is quite likely the work will have to be released from the chuck before the tool can be extracted.

The obvious answer is not to let the heat build up to the point of seizure; there are several things that can be done to help. Once again it must be said, ensure the cutting edge of the tool is sharp and at centre height, if not it will be rubbing on the work instead of cutting and there will be extra friction and as a result extra heat. Make sure that the machine is running at an appropriate speed, if it is going to fast the extra revolutions will create heat and if it is going too slow that can cause the cutting edge to rub the work away rather than cutting it. If cutting fluid is appropriate

Parting off using the rear tool post, it is advisable when parting off mild steel to use plenty of coolant and to ensure the lathe is running at the correct speed.

to the metal that is being machined apply it liberally, if not then don't try to help matters by applying something that is not appropriate. The operation can also be kept from overheating by not winding the tool in too fast, however a steady pressure should be maintained.

The other main cause of failure is a build up of swarf in the groove, eventually a piece gets between the cutting edge and the work and sticks, if heat is also a problem it may even weld itself in position. The cause will probably be that the tool is being fed in too quickly and feeding more slowly and removing swarf from the groove that is being made can prevent the effect. Cutting fluids will wash away some swarf but it may be necessary to physically remove some. The longer swarf is allowed to remain the greater the amount of heat that will be created and the less the tool will cut so the more it rubs. Particular care is needed when machining aluminium, because of its low melting point the metal is inclined to build

It is possible to use a parting off tool for making narrow grooves as well as for cutting off. Providing the tool does not have too much overhangs it can be traversed along the work with safety.

A parting off tool is here being used to make a small crankshaft, the grooves that will ultimately be machined to become journals have been made and the parting off tool is used to finish them off.

up and weld itself to the tool, and the only way to remove the build up is to remove the tool and break the swarf off it. It is quite a vicious circle and aluminium must be kept cool at any price. A constant supply of coolant such as white spirit or paraffin changes the pattern of swarf from a continuous chip to a powdery form and stops the over heating.

The reason for parting off to be so disliked is often due to problems created by the machine operators themselves. It is of the utmost importance that before starting attention is paid to a number of things. Most important is the placing of the tool itself, there should be as little overhang as possible. If a parting tool with adjustable blade that slides in and out is in use, only have sticking out from the body, sufficient blade to go to the depth of cut that is needed. If the cut is to be a deep one consider the idea of having a short length of blade protruding and at some stage extending that amount. Providing the tool has been set up square in the first place the extended tool should slide directly into the groove that has already been made.

For all machining operations the slides should be properly adjusted and attention paid to ensure that tools are firm in the tool post, precautions that are even more vital when parting off.

Enough cannot be said about tool height, if the cutting edge is above centre height it will not be cutting and a lower part of the blade will be rubbing the material away. Imagine the amount of frictional heat that will cause. If it is below the rotation of the lathe will tend to push it down under the work and therefore try and lift the tool up, this can result in the work trying to ride on top of the blade. One of the most common problems is that as one nears the end of the cut the blade rides over the work, frequently ruining it.

Even using the above methods parting off from the normal tool post can be difficult. The rotation of the machine is attempting to force the cutting edge of the blade down all the time. The blade of course doing the opposite and trying to lift the metal up, this means it is pushing against the bearing so pressure is being applied in two directions. No matter how good the lathe there is

A slightly different parting off tool, designed to be held at the side of a tool post.

always a going to be a tiny amount of play, sufficient to create unwanted problems. Setting the tool to the rear of the machine, and to set the parting tool upside down can reduce the problem of lift in the bearings. It is necessary to put it upside down so that the work is rotating towards it; otherwise the tool would do not more than rub against the metal. With the tool upside down it is now pushing the work down towards the lathe bed where there is greater rigidity, there is much less likely to be any flexing of the bottom bearing than there is of the top.

In order to place the tool upside down at the rear of the machine another tool post

is generally required and it is possible to buy rear tool posts if one so wishes. Making one is the simplest of matters and by doing so at least one gets a piece of equipment that is tailor made to ones own needs. Putting two tool posts on the saddle of most lathes will result in very limited space between them, which can make working inconvenient. Most people therefore make a short extension on which to mount the rear tool post. The same precautions need to be taken when parting off from the rear as from the front but the better stability that has been obtained will immediately be felt.

In the case of holders that have the blade set to one side, turning them upside

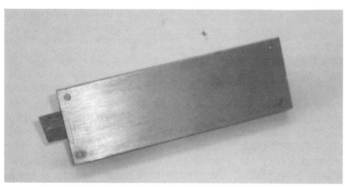

A parting off tool made from a high-speed hacksaw blade. The holder provides plenty of support and the narrow blade can be particularly useful when working on small items or making narrow grooves.

down is not always an option as doing so reverses the position of the blade and sets the cutting edge further away from the chuck than is desirable. It is worth investing in a suitable holder that will bring the blade close as to the chuck as possible.

For all parting off operations the tool should be wound in to the work, slowly and once cutting commences a steady light pressure maintained throughout the operation. Do not hurry, but at the same time do not dawdle either, it is a cutting action that is required not a rubbing one. The recommended way is that a steady pressure should be applied and the movement maintained until complete.

There are always some people that like to be different and do things their way and if something works for them, why not? When it comes to parting off there are those who like to start by making a small groove and then withdrawing the tool. The tool is then moved about five thousandths of an inch or a quarter of a millimetre to one side and another short groove made. The tool is again withdrawn and moved to its original position for another short cut and so on. The idea of course is to prevent any binding up and from that point of view it is quite successful. Sometimes a thin blade is likely to move sideways when it comes to one of the ridges that have been formed. This will not harm the tool, the work or the machine but it will result in an untidy finish with a series of ridges that may well need to be machined off. While the above idea may result in an untidy finish it is certainly a good one to use if large diameter material is being worked on.

A wider blade is a must for any work of a diameter of over 1-1/2in or 30mm. If it is proposed to use the above idea of making

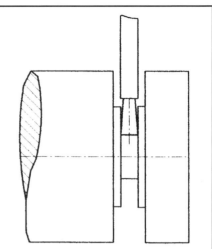

When parting off thick material it is a good idea to make the first cut to partial depth. Move the tool half its width to one side and make a second cut. Move it to the other side of the original cut and repeat the operation. Return to the original position and either finish the cut or repeat the procedure depending on the thickness of the metal

cuts in slightly different positions the sideways movement each time should be on alternate sides of the groove, taking care that every third time the tool is back in its original alignment. The final cut is taken on the alignment of the side on the part to be removed and if the job has been done properly, there should be no ridges.

Many people have over the years developed their own method of parting off and if a suitable way to do so has been found it is best to stick to it and not experiment in the hope of finding something better.

As well as parting off the parting tool is very useful for cutting narrow grooves in work and fortunately because such grooves are rarely very deep the overheating problem does not exist. If a radius is required at the bottom of the groove this can be ground on the tool and the normal edge replaced when it is required for its more usual operations.

A lot has been said about the width of the tool and other than when heavy work is required this may matter very little, however when machining grooves commercially made blades may be too wide and it will be necessary to have a tool that will cut the narrow grooves required. The obvious answer seems to be to grind the blade until it is narrower. It is very difficult to reduce an already narrow piece of high-speed steel to a yet narrower profile. The end result more often than not will be a series of steps, which may or may not prevent the blade from being capable of use. There is a way to produce ones own very narrow high speed parting off blades, with a minimum of trouble and expense and that is to use a discarded hack saw blade.

We throw worn out hacksaw blades away by the dozen and yet even though the teeth are worn the steel from which they are made is still very good. All that is required is to make a suitable holder, grind the teeth that remain on the blade off, sharpen the cutting edge and hey presto! We have a perfectly useful parting off tool. True it will not like cutting through 1in or 25mm thick steel, but it will certainly part off steel up to a diameter of around 1/2in or 12mm diameter and even thicker brass, it will also cut nice narrow grooves.

To sum up, if reasonable care is taken and thought given to the process parting off is not all that difficult. Most problems that occur are the result of blunt tools set to the wrong height and slackness in the slide ways of the machine.

Chapter 11

Hole boring

So far all the machining operations that have been described relate to working either on the outer surface of material or across its face. A good percentage of machining operations are involved with working inside, in other words drilling and boring holes and recesses. Every reader will no doubt be familiar with drills, or drill bits as they are frequently referred to and now it is time to consider their use on a lathe. The main difference is that when we drill on the bench, the drill is rotated, on a lathe the drill is stationary and the work revolves. In most instances this will make little or no difference, except that most lathes will be considerably more powerful than a drilling machine, with the result that if one gets careless, or a mishap takes place there is much more chance of the drill breaking. Fortunately the inbuilt strength and accuracy of the machine means that it is harder to be careless.

It is usual when working on a drilling machine that having carefully marked off the place where a hole is to be drilled, to make a mark with a centre punch. On a lathe this also applies to holes being drilled off centre, when drilling a hole centrally in a bar of metal, marking off is not necessary, neither is the centre punch mark.

Centre drills

All holes should be started with a centre drill and that applies whether he job is on a lathe or in a drilling machine. The centre drill should not be taken in too far, the depth of the pilot and just a touch of the countersink section is quite sufficient to ensure a good accurate hole. Centre drills come in a variety of sizes and one with a pilot about three quarters of the diameter of the hole to be drilled is usually about the right size. Not that it is possible to stick to that as a rule as when large holes are being drilled such a centre drill is not available and it will be necessary to use the largest

A centre drill is solid and stable and should always be used to start holes. The photograph not only shows a hole being started in this way, but also an interesting set up. The casting could not be held in a chuck and so was bolted to a piece of steel plate that was then set up in a four-jaw chuck in the normal fashion.

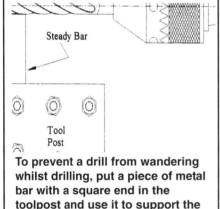

To prevent a drill from wandering whilst drilling, put a piece of metal bar with a square end in the toolpost and use it to support the drill

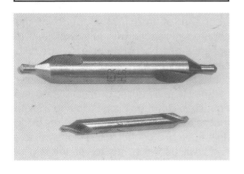

size one happens to have. Most drilling operations can be carried out with the drill held in the tailstock chuck, depending on the size of lathe that is. The chucks on small lathes will in all probability only take drills up to about 6 or 8mm diameter larger models will be capable of holding 5/8in or 16mm.

If working on steel of any description or cast iron, ensure that the drill is really sharp before starting. Set the lathe to a suitable speed, in theory the drill should rotate at the same speed as a bar of metal of the same size would need to. For lathe work it might be beneficial to only run the machine at about two thirds of that speed. Feed the drill slowly into the work, fully withdrawing it at frequent intervals to clear the swarf, which will build up inside the hole if it not cleared. One of the most common causes of broken drills is caused by the swarf not having been cleaned out of the hole regularly enough.

Drills do tend to have a mind of their own and are likely to wander off, resulting in them starting to wobble. This can be prevented by holding a piece of wood against the drill to act as a form of steady, If the drill is allowed to continue to wobble

A pair of centre drills which are available in a variety of sizes. Most are made of high-speed steel, but it is also possible to get harder ones made from cobalt.

Above: A centre drill that has been extended in length to allow access to difficult places. It is simply a case of drilling a mild steel bar and securing the centre drill with an adhesive, retaining compound such as Loctite. Below: The centre drill can be followed by a drill that should accurately follow the line made by the centre drill. The drill will almost certainly wander off line if the centre drill is not used first.

not only will the hole be off centre but it will almost certainly finish up oversize. The use of a coolant is beneficial in drilling operations but only if the metal being drilled is suitable for that type of coolant and one should never under any circumstance be used with cast iron.

Large diameter drills that will not fit in the tailstock chuck are likely to have a morse taper shank and can be held directly in the tailstock. Most of these large drills are quite long and because of this are inclined to wobble when in use. They also require considerable pressure to be applied in order to drive them into the work.

How to prevent a drill wandering by placing a square-ended piece of metal bar in the tool post as a support for the drill

119

Drills that are too large to fit in the chuck can be supported directly in the tailstock mandrel. Usually, a pilot hole is drilled with a smaller drill prior to using the large one.

An answer to these two problems is to use a method of progressive drilling, i.e. starting with smaller diameters and gradually increasing until the final one. Even so care has to be taken to prevent the drill from wobbling. Where possible large diameter holes are better made by boring which gives far superior results to drilling.

Drilling brass, copper and bronze brings different problems, as there is always the chance with these materials that the drill will snatch. When it does this, it jams into the metal and is quite likely to seize up altogether. The problem is particularly prevalent if drills are sharpened before use and also when progressive drilling is taking place. There is a special way of sharpening drills to prevent them from snatching, but it

involves virtually ruining a perfectly good drill. Just as good a result can be obtained by just lightly rubbing a smooth oilstone over the cutting edge before drilling, this will stop the snatching from happening.

Extra care must always be taken on any material when a drill breaks out at the other end of the hole and again the problem is that the drill snatches. If it does the result is likely to be jaggedness around the edge of the hole and in the case of cast iron badly broken edges.

Very small diameter holes and by this is meant holes of less than 1/16in or 1.5mm create problems of their own. One of the main ones being the difficulty in holding them as very few tailstock chucks will close up small enough to do so. Using a pin chuck

There are various types of reamers and the one shown here is a hand reamer designed for use in conjunction with a tap wrench. Reamers will always follow the path of the hole they are used on and cannot be used to correct alignment.

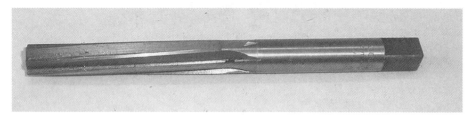

This is a machine reamer and it can be seen that it has a shallower lead into the hole than the hand reamer. It is fitted with a Morse taper to enable it to be fitted into the tailstock.

in the larger one solves that side of things easily enough, but the other problem is with lathe speeds. Small drills of this size require speeds that are impossible to get on many, if not most lathes. It is certainly unlikely that a belt driven single-phase machine will be able to do so. There is little that can be done about it other than to run the lathe as fast as possible and just take a lot of extra care. In particular withdraw the drill to clear the swarf as frequently as possible and do not apply too much pressure with the tailstock.

Reaming

The reason for reaming holes is to get greater accuracy than is possible with a drill alone. Very few drills do cut an accurate hole; it only needs the two cutting edges to be very slightly different in length, or the cutting angle to be very slightly wrong for the hole to be either oversize or slightly oval. If a reamer is used to remove the last bit of metal from the hole it will correct this, but what it will not do is correct faulty alignment as a reamer always follows the pilot hole. Exactly how much metal a reamer should remove is hard to say, other than that it should be as little as possible. The actual amount will depend on the metal involved and the diameter of the hole. A large hole of say 1in or 25mm could easily be drilled 1/64in or 0.4mm undersize, while a 1/8in or 3mm diameter hole would want no more than 0.006in or 0.15mm to be removed.

Reamers can be held in the tailstock chuck or if too large for that it will be found that they usually have a small hole in the end and a hardened centre can be located in it. The reamer can be then held with a

A number two Morse taper reamer, with a No. 2 taper for fitting it in the machine. Available in most Morse sizes, these reamers are generally used for cleaning out tapers, but can also be used for correct alignment after boring.

Hand reamers are not well suited for machine use. Like all reamers they have a tendency to snatch and stick and the parallel shank gives insufficient grip to prevent this from happening. They are best used as shown, with a tap wrench and supported on the tailstock centre to give correct alignment.

tap wrench while the lathe is rotated by hand. A lot of pressure is involved in reaming and it is essential that the reamer be fed into the hole very slowly. There are a large number of cutting edges, all creating friction in addition to cutting and if care is not taken when feeding the reamer in it will bind in the hole. Frequently a reamer that has stuck in this way is very difficult to remove. The obvious way to do so is by reversing the direction of rotation of the

lathe. The reamer should be withdrawn from the hole at frequent intervals, to clear the swarf and this should stop it from binding in the hole.

It is possible to ream with the lathe under power but because of the difficulties referred to above it is often better to rotate the machine by hand. A reamer that has stuck when used in a machine under power can be almost impossible to remove, short of applying a hammer to the end. The

Using a small boring bar. Note that the bar is in a special holder that allows more pressure to be applied when tightening the bar in the tool post than would be possible if the boring bar went directly into the tool post.

A larger boring bar that fits directly in the tool post, the greater surface area of the square section provides an ample surface for the bolts in the tool post to grip on.

resulting hole is far from smooth and frequently it means scrapping the job. If a drilled hole has wandered off centre, do not expect a reamer to align it properly. Reamers will always follow the line of the hole they are being used on, whether it is concentric or not.

Boring

By far the best way of making a hole when using a lathe is to bore it. Done carefully the result can be a fine finish and absolute accuracy, even so it is a task that requires care and patience. Boring tools are virtually the same as ordinary lathe tools, except that they have a long shank on them, they are made of the same materials as ordinary tools. Such differences as they may have from ordinary turning tools are in the methods of holding some of them in the tool post.

It is necessary to start with a drilled hole

Boring a cylinder casting mounted in a keats angle plate, the use of which makes it easier to set up accurately than if it was bolted directly to the face plate.

123

Above: The use of a single tool post like this allows better vision of what is happening during boring operation, when a larger tool post may well be in the line of vision.*Below:* Close-up view of the tipped tool, showing the shape of the tip. The front rake on such a tool may mean that it will only be suitable for boring comparatively large diameters.

in virtually all instances, although a hole can be bored without first drilling it is a laborious process and involves working from a centre point outwards in a series of very fine cuts. This is not to be recommended unless it is absolutely essential that the work must be done in that way.

Once the pilot hole has been drilled the procedure for boring the hole basically follows the same pattern as ordinary machining operations. There are some slight variations, the first being that as a rule the boring bar will tend to flex during cutting, moving very slightly towards the lathe centre. It is a perfectly natural thing to happen; as has already been pointed out there is always the chance of a tool flexing if it has too much overhang. The very nature of a boring bar means that is will have a great deal more overhang than a normal turning tool and some flexibility is inevitable. The obvious answer is to keep the length of the boring bar as short as possible but this is not an easy thing to do as unless one has an almost inexhaustible supply it is necessary that the tools we do have are long enough for any job.

Assuming therefore that one is stuck

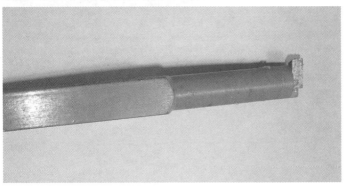

A cemented tip tool. This type of tool is particularly useful for boring castings where it is desirable to break through the surface in a single cut.

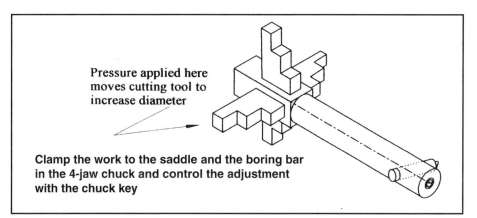

Pressure applied here moves cutting tool to increase diameter

Clamp the work to the saddle and the boring bar in the 4-jaw chuck and control the adjustment with the chuck key

with a boring bar of a particular length another solution has to be found. The obvious one is to run the tool in the hole as many times as necessary for it to remove the right amount of metal so that there is no longer any pressure causing the flexing. It is possible to wind the tool back, with the lathe rotating and it will cut as it comes back. The only snag with that idea is that even at the start of the operation the tool will have flexed and running it back may allow the tool to spring back leaving the hole slightly oversize. To prevent this it is best either to wind the tool back while the work is stationary or to actually move the cutting edge away from the work, withdraw the tool and commence the cut again from the original cross slide setting.

Whichever way it is done, the cutting must be repeated until the bore is the correct size. To ensure it is the correct size throughout its length it will be necessary to measure it. A short hole can be measured with vernier or digital callipers reasonably successfully. Better still if an internal micrometer is available that will be even more accurate. To measure longer holes one either needs a proper bore gauge, or

alternatively it is possible to use inside callipers. A final and highly accurate solution is to make a plug gauge. This is not so difficult as it may at first appear, as it is simply a case of machining a length of metal to the exact diameter of the bore. It is pushed into the bore and any tight or loose spots become obvious. Of course unless one possesses two lathes it will have to be made before the boring operation commences. Other wise it would be necessary to remove the work in order to

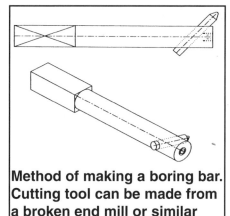

Method of making a boring bar. Cutting tool can be made from a broken end mill or similar

125

Boring between centres, the work is mounted on the cross slide and it is an extremely accurate way of working. Unlike the boring tool held in the tool post the between centre one cannot flex as it is held rigidly between centres.

make the gauge, and it would then be impossible to set the work back accurately again. Of course if the bore is really long, it is probably not going to be practical to machine something to fit the whole length. The gauge therefore will have to be drilled and tapped and a length of thinner bar from stock material fitted to it to act as a handle.

In order to reduce the length of a boring bar a commonly used method is to make a small vee block and rest the boring bar on it, using the clamping screws of the tool post to hold the bar in position. It is very effective and a quite substantial set up is the result. The idea can be used on very long holes as well as short ones, for roughing out purposes the length of the bar can be adjusted during the actual boring operation.

Steadies

It is advisable to use a steady to support work that is being bored, if it is possible to

A between centres boring bar consisting of a bar with centres at each end, cross-drilled to hold a suitable tool that is held in place with a grub screw. A peg that will mesh with the lathe catch plate is required for driving and in this case a long cap screw has been used. Adjustment of the tool is measured with feeler gauges.

do so, remembering that just because the cutting operation is taking place on the inside of the metal rather than the outside, the same amount of pressure is being applied. It is not of course always practical to do so as many boring operations take place on castings, where no suitable outer surface is available on which a steady can be located.

Boring on the saddle

By far the most accurate method of boring is to clamp the work on the saddle and use the boring tool either from the chuck or between centres. An ordinary square shanked tool can be placed in the four-jaw chuck and the diameter of its movement controlled by adjustment of the jaws. It is a useful way of carrying out the machining where the bore is blind and adjustment of the tool is comparatively easy. The real advantage of using the saddle comes when the bore is not blind and the tailstock centre can support the boring bar. The other end may either be held in the three-jaw chuck or supported by a centre and driven with a catch plate.

The boring bar for this sort of work is of course entirely different to those that are mounted in the tool post. A solid bar is cross-drilled and a tool is inserted in the hole and secured in place with a grub screw, Adjustment of the tool is by releasing the screw and moving the tool. To obtain the necessary fine adjustment a depth micrometer or vernier type calliper can be used. When the bar is of a substantial diameter there is sometimes an adjustment screw that raises and lowers the tool, which is then secured by the grub screw. The system is particularly useful for boring out cylinder castings where it is essential that the bore is absolutely parallel.

Cross drilling

In addition to the more usual drilling and boring operations the lathe is an excellent

Below left: the depth of holes are measured with a depth gauge. The one shown is home made by simply drilling a piece of square steel bar and pushing a length of rod through it. A hole is cross-drilled and tapped and a screw used to hold the rod in place when it has reached the bottom of the hole. Below right: another depth gauge and again homemade. This one is spring-loaded and so does not have a screw with which to retain its position. It is also graduated in 3-mm divisions as an aid to measurement.

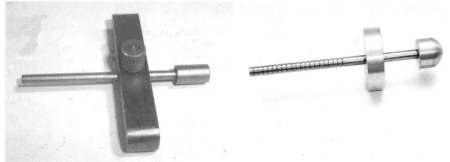

tool for drilling accurately across a round bar. Cross drilling a bar is not the easiest of tasks and normally requires either very careful lining up and then securing the work down so it does not move or alternatively using a specially made jig. All that is required to do the job accurately and quickly on a lathe is a specially adapted centre. Once one has this it is easy to cross drill, using the inbuilt accuracy of the lathe to do so. The centre has a vee cut in the end in which the work rests and by putting the necessary centre drill and drill in the headstock, either using the three jaw or tailstock drilling chuck, the hole that is made will be located exactly through the centre.

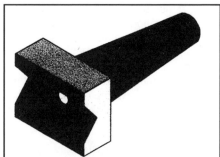

A morse taper with a vee block can be used for cross-drilling round bar. The centre is fitted into the tailstock and the round bar clamped in the vee before drilling in the usual way from the headstock

Chapter 12

Threading

It is inevitable that at some stage or another the lathe will be used for cutting threads, either internal or external. How this is done will probably depend on what the thread is that is being cut. Small threads can usually be cut with taps and dies. Larger diameter threads will call for the work to be done with a single point tool using the facilities offered by the lathe. It will also depend to some extent on the metal that is being threaded as well as the diameter; it is obviously easier to work on a soft material such as brass than it is on something like silver steel, which is quite hard. Each individual will have his or her own idea on the diameter of material that they can safely deal with using taps and dies. There are plenty of tables available that give the drill sizes of holes required for making an internal thread. Threads are usually given as a percentage of the thread depth, a hundred percent would mean that there would be no space between the bottom of

the tapped hole and the peak of the thread as cut by the die. This is rarely acceptable as to tighten or loosen a hundred percent thread is extremely difficult and generally speaking an engagement of around eighty percent will result in a good leak proof fit. Seventy percent will be a good average for threads that will need to be tightened, loosened or regularly adjusted. Like most things it is a case of 'horses for courses' and a thread for something like a lead screw will need a deeper engagement.

Its is all very well knowing the size and shape of thread that is required, but suppose there is a need to duplicate the thread on something? This most commonly occurs with bolts, how then do we determine what the thread is? Well one way is to measure the outside diameter and then count the number of threads per inch. Of there is no need to count a whole inch of thread, just say a quarter and then multiply the number by four. This still does not give

A set of thread gauges. These save a lot of time, the alternative being to count the number of threads.

us the angle of the thread but as the number of threads per inch is only in the odd instance duplicated the idea works quite well. In the case of a metric thread it will be necessary to count the number of threads over a given distance. For example if we measure 10mm along the master and there are ten threads, then the pitch is 1mm. An easier way is to use a thread gauge, something that can be purchased very cheaply and consists of numerous leaves, the edges of which have the profile of teeth cut into them. It is simply a case of checking until the profile that matches the thread in question is found. Gauges are available in imperial and metric ranges and are well worth while investing in.

Assuming at this stage that we are talking of small diameter external threads, they will be made with a die and if possible the die should be held in a holder in the tailstock. Tailstock die holders are readily available and in addition, like so many of these small tools, easily made and doing so is a good way of improving ones skills on the machine. There have been numerous designs published for their construction. Threading a piece of rod requires a great deal more power than one would expect and if threading material over

say 6mm or ¼in the die holder should be fitted with some form of tommy bar in order to assist the process. As diameters get larger so it becomes less likely that a holder located in the tailstock with a morse taper will be secure enough and the holder plus the morse taper will almost certainly start to rotate, the movement can be arrested if a tommy bar is fitted.

Dies that are used in Britain are mostly of the split variety, which means a gap has been made between the outside and the thread. The majority of other countries do not use this system and dies are not split at all. There are three screws to hold the die, which has three recesses for them to fit in. The centre one of these is actually in the split and offers some adjustment of the diameters of the thread produced, so if the split is opened as wide as possible a larger diameter thread will be produced than if it is closed up as far as it will go. The idea is to start making the thread with the split at its maximum width, then repeat the operation, having tightened the two screws on either side of the split to close it up and take another cut with it in that position. What one needs to be very careful when doing this to ensure that the die is not tightened so much that a loose thread is formed.

Tailstock Die Holders. The picture shows three different size of holder that all fit the same morse taper. As threads get larger it becomes necessary to use a tommy bar in order to give extra pressure. The tommy bar can be permanently fitted or made as a screw in fitting.

If the pressure required threading with a die using the tailstock die holder is too much for the morse taper, an ordinary die holder as used for bench work can be used. One trouble with this type of die holder is that it can all too easily move out of line and the thread that is formed is then at an angle. To avoid this a die holder should be supported from behind, with the tailstock barrel and as long as the die is at ninety degrees to that the thread will be square. The use of a thread-cutting compound is strongly advised as not only will it ease the workload, making the job easier, but the end result will also be improved.

Internal threading using taps

The first requisite is to drill a hole for the tap and it is as well when using imperial threads to use a published table of tapping sizes. If one is not available then the root diameter of the thread will give a hundred percent engagement and it is possible to work the percentage from that. In the case

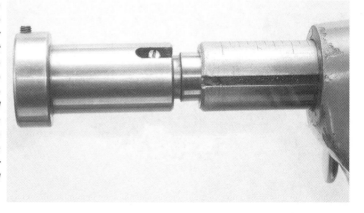

One of the tailstock die holder shown in the previous picture ready for use. Not the pin that runs in a slot in the holder, this allows the thread to be made under power and when the die has travelled a certain distance the pin slips preventing any further rotation, until adjustments are made.

Another type of tailstock die holder, this one is double ended with space for a different sized die at each end. It is not fitted with a tommy bar but has heavy knurling to enable a firm grip to be taken to prevent it rotating.

of metric threads although there are charts available a good rough guide is to subtract the pitch from the diameter and use a drill of that size. Threads of fittings for gas and water need a greater depth of engagement than normal types used just for holding items together.

There are a number of types of tap and each type is generally sub-divided into three types, these are taper, second and plug, the latter frequently referred to as bottoming. The names more or less speak for themselves, the first has a tapered end to allow it to work its way into the hole, the second has a slightly smaller taper and the plug virtually no taper at all. They should be used in that order, although many people dispense with the use of the second tap and work with just a taper and plug, something that seems to make little difference to the end result. When tapping steel of any sort it is wise to use thread cutting compound to lubricate things. These compounds vary considerably, some are thin and runny, others are thick and almost sticky; the type used is a matter of preference for the individual. In instances where for some reason or another, a

A very small diameter rod held in the tailstock chuck fitted o the lathe mandrel, being threaded with a die held in the tailstock.

Threading with a tailstock die holder, this ensures accuracy and a true thread.

suitable tapping compound is not available, a number of substances can be pressed into use, typical examples being ordinary cutting oil, any form of grease or neat washing up liquid. The main aim is to lubricate the work, but most tapping compounds are also designed to retain the swarf in such a way that it does not cause the tap to bind.

Tap holding

It is important that the tap is held in a way that is absolutely truly in line with the hole to be threaded. This may sound both common sense and easy, common sense, yes, easy not perhaps quite so easy as it may seem. Whether the work is being done on a lathe or a bench without it being noticeable the tap can be wound in at an angle. Not necessarily a very steep angle as that will be obvious, but one or two degrees can spell disaster. Not only will the thread not be square but also more taps are broken as a result of them being taken in at an angle than any other way. When tapping on the lathe unless one is very careless getting the threads square is comparatively easy, unlike on a bench

Tapping piece of work with the tap held in the tailstock chuck.

133

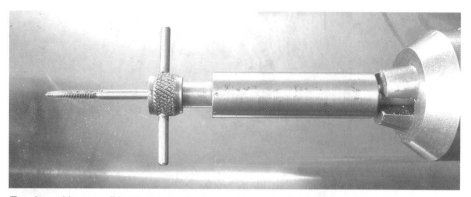

Tapping with a small home made device that is more delicate than the tailstock chuck and therefore more appropriate to small taps. The section holding the tap is free to slide along the section held in the tailstock, meaning that taps are less likely to break.

where unless a guide is used it is a case of judging matters by eye.

The most obvious way to hold a tap so that it in line with the hole is to set it in the tailstock chuck and no doubt that it the way most of us will work. It is possible to obtain special tap holders that have a ratchet mechanism operating a form of clutch, if the tap sticks the mechanism allows it to rotate with the work, thus preventing it from breaking. Such devices are very expensive and are hardly worthwhile unless it is proposed to carry out a very large number of tapping operations. Taps sticking in the work are a perpetual hazard and as a rule if they are held in the tailstock chuck they will rotate with the work and slip in the chuck jaws, thus providing a safety measure.

The actual tapping operation requires a great deal of care and patience, it is

The parts of the tapping device shown in the previous picture, it is easy and quick to make, Most very small taps have a standard shank that will fit the device, it is only a matter of a few minutes to make extra holders for different sized taps.

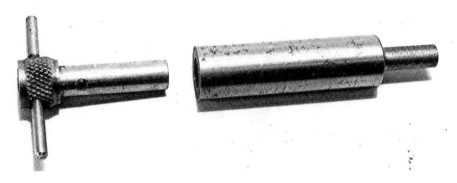

Sometimes it is difficult to stop a tap from rotating in the tailstock chuck. An alternative is to use a tap wrench supported on a centre in the tailstock.

essential that after every rotation or two the tap must be fully unwound and if necessary cleaned to remove any swarf that may be embedded in it. The deeper it goes the more frequently it must be withdrawn. Trying to wind a tap straight into the work is asking for trouble unless it has a very low engagement.

If a tap becomes stuck in the work and it is something that happens to everyone, a great deal of patience is required to remove it. The first necessity is that if the lathe is under power, switch it off. Assuming the tap in held in the tailstock chuck; only remove it from the chuck as a last resort. Start by trying to turn the machine by hand in the reverse direction, it might work, but frequently will not. It then becomes a case

When screw cutting unless the lathe is fitted with a gearbox it is necessary to set up a train of gears like this in order to obtain the required pitch.

135

of trying to alternately move the lathe by hand in each direction, no undue power should be applied and only the tiniest of movements made. This very gentle rocking movement will usually do the trick but requires a great deal of patience. The fact that the tap seizes is almost inevitably caused by a tiny piece of swarf becoming lodged, between the threads in the work and those in the tap, and once this has been freed all will be well. Once the tap has been freed, wind it right out and clean out the flutes and the threads thoroughly. An old toothbrush and a drop of white spirit do this quite nicely. The tap should then be lubricated once more and operations resumed, remembering at all times to take care. Once a tap has broken in the work it is very difficult to remove the broken part and may well result in the work having to be scrapped.

The diameter of thread that it is possible to make using the tailstock chuck is quite limited; an exact diameter cannot be given, as it will depend on many facets. Whatever that maybe there will almost certainly be a time when it will be necessary to have a more secure hold on the tap in order to be successful. We use a tap wrench for normal tapping operations at the bench and so now it is necessary to resort to one for working on the lathe. We are faced with the problem referred to earlier of keeping a correct alignment. Fortunately most taps and certainly all high quality British ones will have a centre hole at the top of the shank and this can be used to keep the alignment correct. The tap is fixed in the wrench and the tailstock fitted with a hard centre is brought up to bear in the centre hole. It is now possible to either rotate the lathe by hand or rotate the tap wrench if there is sufficient room. The same method of applying a turn or two and then winding back must be applied and it is necessary to keep winding the tailstock in so that the centre remains in contact with the tap at all times.

If it so happens that the tap does not have a centre hole in the shank all is not lost. Make sure that the end does not protrude past the tap wrench and put a small piece of flat metal against the end of the tailstock mandrel and use that to keep the tap in line. The tap wrench should have a small flat section around the hole that accepts the tap and the metal placed across the mandrel should be flush with that flat part of the wrench and it will keep things lined up. It is not quite as easy as it sounds as one hand is required to rotate the lathe, another to hold the flat metal against tailstock and tap wrench and a third to rotate the tailstock. In fact what happens is the piece of metal keeps slipping out and has to be retrieved before proceeding, and it is nowhere as difficult as it sounds.

Readers will have noted that it has been stressed that the lathe should be rotated by hand during threading operations, using taps and dies and that is by far the best thing to do. When tapping larger holes it may be possible to use power, but the speed should be as low as possible without using back gear. If the thread has a very low engagement then it will be easier to use power. Unless the thread is very tight it is as a rule quite safe to use power to reverse the tap out.

The best way to rotate a lathe by hand is to fit a handle fitted at the rear of the mandrel, the alternatives are to either use the chuck or pull it round with the drive belt. It is possible that if it is rotated by pulling

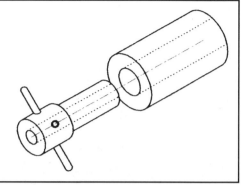

A useful tapping aid. The part on the left holds the tap which is secured by a grub screw. The right hand part is held in the tailstock chuck. The tap is rotated by hand and is kept in alignment by the device

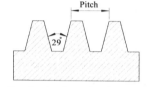

Drawing showing angles of acme thread. Dimensions for this type of thread have been laid down in BS 1104 of 1957

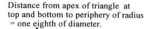

Distance from apex of triangle at top and bottom to periphery of radius = one eighth of diameter.

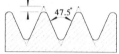

Angles of BA threads

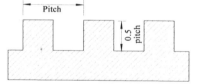

Formation of the square thread.

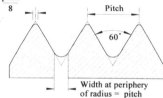

Width at periphery of radius = $\frac{pitch}{8}$

Width at periphery of radius = $\frac{pitch}{4}$

This thread form applies to both metric and both unified thread pitches

Specifications for various threads and the angles to which the tools must be ground in order to cut them. The Acme and Square threads above are used purely for transmitting power in the form of lead screws, etc. The others are standards for nuts and bolts

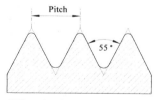

Whitworth and BSF thread form, same as for BA except angle at 55 degrees

on the chuck and the chuck is screwed to the mandrel it could cause it to unscrew, although this is not very likely. Any arrangement for a handle will almost certainly have to be home made, but such a device is well worthwhile making as it is something that will find frequent use.

Cutting threads using the lathe

Arguably one of the most satisfying tasks on a lathe is cutting a thread, using a single point tool. The first thing one needs to know is the complete specification of the thread to be cut and for this a table is required. There are several measurements that need to be known and the first is what type of thread we are about to cut. There are a number of thread types and although gradually the British ones are being phased out in favour of the metric system, it will be many years before they disappear altogether. Many of the drawings used by model engineers were made many years ago and the threads they specify are Imperial, also anyone wishing to preserve a British motor vehicle may very well

discover that the threads are to an Imperial standard or standards. Later vehicles had threads of the American standard, so there is and will be for years to come a wide variety that it is necessary to have some knowledge of.

We must therefore ask ourselves what we need to know about the required thread and the answer is the outside diameter and the core diameter, which in other words is the depth to which the thread has to be cut. In addition it is necessary to know the angle of the sides and the pitch. The pitch means the number of threads over a given distance and for British and American threads it is specified as so many threads per inch. When it comes to the metric system the pitch is given as the distance between each thread. So if a British or American screw is shown as the outside diameter times 18tpi or eighteen threads per inch, a similar metric one would be shown as the outside diameter times 1.4. To confuse matters a little more, as if it wasn't confusing enough already, some American and some British threads are given as a number. The British

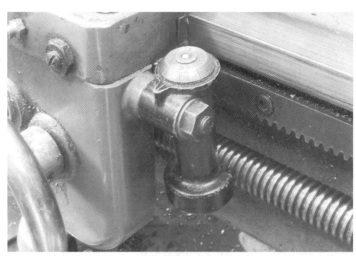

A thread dial indicator, something that will be required for cutting threads whether the lathe has a gear box or not.

Association or BA system as it is generally referred to has thread numbers from zero, which is the largest to number 25. Whether or not anyone has ever seen a 25BA thread is open to conjecture and matters not; the important thing is that the lower the number the larger the thread. The Americans number their threads in the opposite direction the smaller the number the smaller the thread. Fortunately it is very rare, although not completely unknown for anyone to cut one of these numbered threads. They are small enough to be made with a die, so on the whole there is no need to worry too much about them. All that is required is to know what their outside diameter is and the size of the drill required when tapping them.

Of course it was never going to end quite as easily as that and yet another range of threads raises its head, it is known as the British Standard Pipe and in spite of its name it is used in America and all over the European continent as well. The difference in the pipe series and all the others is that the thread is specified by the internal

The screw cutting gauge has shapes cut out of it to match all threads, It not only acts as a gauge to getting the correct angle on the tool but is also a valuable guide to setting up for screw cutting operations.

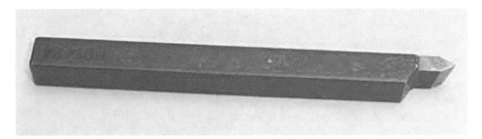

A lathe tool ground to the correct angle for screw cutting, the angle should always be checked against a gauge such as the one in the previous picture.

diameter size of the pipe and not as in the other cases by its outside diameter.

In order to cut a thread by machine it is necessary to set up a suitable feed for the carriage that will ensure the travel of the tool is identical to the pitch of the screw that is being machined. If the lathe has a gearbox then it is simply a case of selection. If it is fitted with change wheels, it becomes a case of setting up a suitable train. This means driving from the mandrel to the feed screw and obtaining the correct pitch by meshing the appropriate gears. The gears will be running on spindles located on a quadrant at the rear of the lathe, the spindles can be adjusted by simply undoing a nut and sliding them to the appropriate position, the quadrant is also adjustable with a couple of bolts. In order to set up some gear trains it will be necessary to use gears that do not actually carry out any form of reduction, they just simply act as idlers, and giving the required space for two other wheels to mesh.

The angle of the thread varies and for the sake of convenience we will for the purpose of this book stick to sixty degrees. It is the angle used by both the metric system and the American ones and so it is a convenient figure to work with. The first

requirement, apart from knowledge of the thread is a suitable tool with which to do the job, which means a pointed tool, ground to exactly the angle of the thread. Once it would have been necessary for the operator to make this to shape for him or herself, nowadays for many threads it is possible to buy a tipped tool that is ready made for the job.

Also required is a gauge for setting up, unlike the gauges referred to earlier this is a single steel plate with a series of cut outs that match the shape of the tool. In use it is laid flat alongside the work piece and the tool is moved into it and the angle of the tool can then be adjusted so that it fits exactly in to the gauge. If when this is done the top slide is set to zero, when the tool is wound into the work it will cut on the point and both edges, rather as a parting tool does. It is actually far better to set the top-slide to half the angle of the thread, so if we are cutting at sixty-degrees the top slide needs to be at an angle of thirty degrees. This way the tool will cut only on the leading edge, giving a far better finish and a smoother cut.

Assuming everything has been set up, the saddle is engaged by the self-act lever in the same way that it is engaged for fine

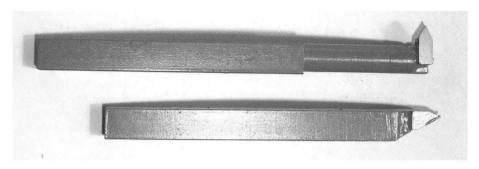

A pair of tipped screw cutting tools, internal and external. Being made especially for the job there should be no need to check that the angle is correct, however they will be specific to a particular thread angle, in this case 60 degrees.

cutting. If we stop to think about it the fine feed arrangement is really a screw cutting operation, except that the threads are so closely spaced together as to merge with each other. One big difference is that if the self act is disengaged and the tool brought back to where it started when the lathe is set for fine feed, it matters not what the position is when it is re-engaged. When cutting a screw it is not quite that simple. Engage the lever at any old position and the tool is likely to restart somewhere between the threads that have already been

cut and make a right mess of things. Therefore it is necessary to know the exact position at which the lever should be engaged. The device that shows the way is known as a thread dial indicator basically a worm wheel with a thick spindle on the top of which is a graduated dial. The pitch of the worm wheel depends on that of the lead screw. During normal machining operations the indicator is not in engagement with the lead screw can be moved so that the worm connects with the lead screw, it is engaged by undoing a nut

For external screw cutting that gauge should be held flat against the work in such a way that the cutting tool point can be lined up exactly with the required vee that matches the thread.

When internal screw cutting, the gauge should be placed flush against the chuck face to ensure it is square and the tool inserted in the vee as in the previous picture.

and pushing it over. Once it is in engagement the dial rotates, being driven by the lead screw. Assuming that the carriage itself is in engagement with the lead screw it will move at the rate that is dictated by the gear train. Providing the dial is relocated at the same position every time the carriage is engaged, cutting will commence at exactly the same place as previously. In fact depending on the thread that is being cut it could be possible to engage at several places, but as that will depend on the lathe lead screw and indicator, readers will either have to use the

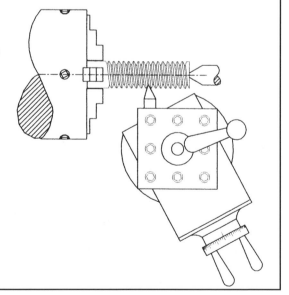

The set up for screwcutting, top slide set to required angle and tool post at 90 degrees. This means that the tool is cutting on one face only. The other side of the tool is in light contact giving a polishing effect. If the top slide is wound in a 90 degrees, cutting will be more difficult and the finish will be rough

An example of screw cutting in operation.
Photograph courtesy of Proxxon Tools

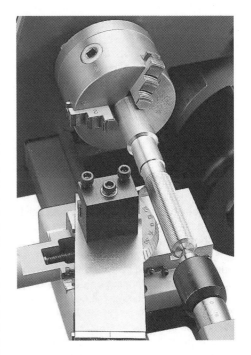

same number or sort the divisions out by reference to the lathe handbook.

One thing that is certain is that if the engagement is at exactly the same place each time the tool will engage correctly. This is a convenient fact for anyone that does not happen to have a thread dial indicator as it means that it is possible to find the right place anyway. Simply put a chalk mark by a tooth on the first gear in the train and a mark on the lathe itself, adjacent to it and if the self act is engaged each time the two marks line up the tool will start cutting at the same position.

The first cut that is taken should be very light; in fact it need hardly scratch the surface of the metal. This is to give one the opportunity to check that the machine is actually cutting the pitch that it was assumed to have been set for. Once satisfied that it so doing, actual thread cutting can commence. Unless there is a good reason why one should not do so, it is a good idea to cut a small groove at the position where the thread is to end. This will allow the tool to run out. Otherwise it can be difficult to find the exact finishing point.

The tool should always be fully withdrawn at the end of each cut; it is not unknown for people to leave it engaged and simply reverse the direction of the lathe. The only result in that case can be a very sloppy thread, the backlash will result in the tool returning in a slightly different position which is not what is required at all.

It is advisable to run the lathe at a slow speed when cutting a thread, it is very easy to misjudge the position at which the tool

has to be withdrawn and if the lathe is rotating at a fast speed there is much less margin for error than when it is going slowly. It is also a good idea to use plenty of cutting fluid, it will help to clear away the swarf, which can become something of a nightmare when screw cutting.

So far it has all been about cutting an external thread, making an internal one is a little more difficult. First of all lining up the cutting tool is not quite as straightforward, it can be set on the actual bar being worked on in the same way as it was for an external thread, but that might be a problem if the work is of a large diameter. It means taking the cutting tool right across to the far side of the material and this may not be practical. The usual

143

way is to put one edge of the gauge against the chuck face and line up on it that way. It is almost essential that a relieving groove be cut at the end of the thread when working internally, it will not be possible to see the end of the cut and although some form of stop, or perhaps just a mark can be arranged it is much safer to run out into a groove.

Many people like to finish a thread by using a chaser. A chaser is a tool that has the outline of the thread cut into it, it is brought, usually by hand up to the thread that is being cut and will improve its finish and accuracy. Chasers can be bought at tool shops; there are various different types. It is necessary to use a different one for each type and size of thread that is being cut, so if one is planning on doing a great deal of thread cutting it will be necessary to buy as many chasers as threads are likely to be made. Chasers are a good idea but assuming the thread that is being cut is a one off, buying a chaser to finish it might be something of an expensive luxury. Sometimes it might be possible to use a tap or die to give that last finishing touch, in fact it is quite common to screw cut a thread purely because too much pressure would need to be applied if using a tap or die, there is no reason whatever why a tap or die should not be used to finish such a thread.

Left-hand threads

Occasionally it is necessary to cut left-hand threads instead of the more normal right hand ones. This does not create any major problem just simply think in opposite terms to the more normal operation. So we need to set the top slide to the same angle but in the opposite direction and run the lead screw in the opposite direction, it is also necessary to alter the clearance angle on the cutting tool, to the opposite side. Those things apart generally speaking making the left-hand thread is no more difficult than making a right-hand one. However one problem that might arise is with the setting over of the top slide as there is a chance that the angle may cause it to interfere with

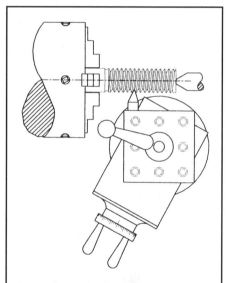

To cut a left hand thread the top slide and tool angle should be reversed because there is a chance that the topslide will foul the chuck

Chapter 13

Taper turning

The first thing one must remember is that it is absolutely essential that the cutting edge of the tool is at exact centre height. If it is not there is every chance the taper will not be true. The type of taper that is likely to be required most frequently is a simple chamfer and this is easily done either by setting the top slide to an angle or alternatively using a tool with a flat cutting face, in other words a form tool and moving it straight into the work.

There are several ways by means of which it is possible to machine longer tapers, the most obvious being to set the top slide over to the required angle and use that. It is a convenient way of machining tapers providing they are not too long. Once the top slide has been moved to the required angle, the tool should be moved to the position that will be the commencement of the taper. The saddle must then be locked to prevent any further

To machine a morse taper, start by drilling a small hole in nff cut of metal and leave it in situ. Next place the pointed end of a morse taper centre in the hole and support the other end with a centre in the tailstock.

Put a flat bar in the tool post, having fist established that the tool post is exactly square with the top slide. Turn the top slide round until the flat bar is flush throughout its length with the side of the taper. Then change the bar for a dial gauge and move that along the taper, using the top slide handle, Check that there is no deviation of movement along the whole of the travel.

parallel movement, as if it is moved it then becomes difficult to reset things to exactly where they were. Remember to position things in such a way that the top slide can travel the whole length of the required taper. It is then a case of using the cross slide to increase the depth and the top slide to machine the taper over its length and return the tool to its original position. A note should therefore be made of the graduation readings for both top and cross slide, prior to starting any machining. This allows one to return the tool to the exact position after a cut along the taper has been made. The

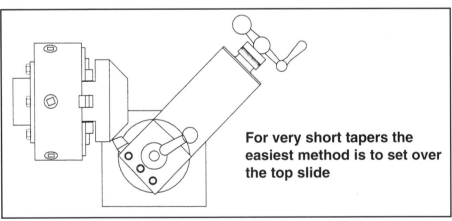

For very short tapers the easiest method is to set over the top slide

cross slide readings will show the exact depth that the tool has been moved.

It is a matter of personal choice whether to machine from the small end to the large one or the other way round although it is easier to machine from the small end as that way the least metal is removed at one pass of the tool. Different people have different preferences as to which way round they like to do the machining. If it is decided to machine from the large end towards the small one, as the tool moves the depth of cut increases and this can be difficult. For example if a Morse Taper is being made, if the tool was to be brought right up to the blank at the start of the machining, by the time it had travelled a comparatively short distance the amount of metal to be removed would be prohibitive. It means that if one definitely wanted to start from the large end inwards it would be necessary to start the cut about two thirds along its length and gradually work back in order to make the machining a practical proposition. Nevertheless there may be situations where machining from the large end is necessary, hopefully these will only occur when the taper required is a very short one and therefore the tool will not need to be taken in to any great depth.

Setting the top slide over by using the graduations is all very well if one is absolutely certain of the angle that is required or if the angle is obtainable via the graduations on the top slide. When setting the top slide angle it must be set to half the angle required, in other words if the piece is to have a four degree taper the top slide

needs to be set to an angle of two degrees. Suppose we wanted to make a number two morse taper, the included angle required is 1.842 degrees. We of course will only be machining with the top slide set to half of that or 0.921 degrees and finding that on a lathe top slide is not very likely. It will be possible to set the slide using trigonometry but many of us will have forgotten all about trigonometry as soon as we left school. In that case we can use a method that is hardly scientific but is easy

Use outside callipers to check maximum and minimum diameter of the taper as a comparison with the original. Make certain that the checks are made at exactly the same place on each taper.

and works very well. Simply put a morse centre between centres, take a metal or hardwood block that is known to have the sides parallel, hold that against the taper and line the top slide up so that it touches the block over the whole of its length. The top slide is now at the correct angle to machine the taper and the idea will work with any morse taper that one wishes to duplicate. It can also be used for shorter

tapers, such as the jacobs type. To make certain that the set over is as accurate as possible, mount a clock gauge on the top slide and move that along the morse taper that is between centres. The clock should not deviate at all from its original setting, if it does then suitable adjustments are required.

In spite of our best efforts getting set up accurately enough to machine morse tapers that will be a perfect fit in a morse taper socket is not quite as easy as it may sound and frequently in spite of an infinite amount of care having been taken when setting up, the taper will not hold securely in the socket A little dodge to cure this is to machine two shallow short lengths out of

It is possible to machine very accurate tapers by putting the work between centres and setting the tailstock over to the required angle. Most tailstocks have the facility for doing this and the picture shows the screw, just above the dovetail on this tailstock that allows it to be set over.

It is not possible to place work between centres in order to machine an inside taper, so it either has to be done by setting the top slide to an angle or by using a taper attachment. To assist accuracy in the case of an inside morse taper special reamers can be purchased, the one shown is for a No. 2 taper.

the taper, generally it will ensure that things become a good fit. At one time commercial manufacturers of taper tooling did this as a regular practice, one that has now died out.

Another way of machining a taper is to set up the bar being worked upon between centres and to set the tailstock over by the required amount to give the correct angle. Most tailstocks have a screw arrangement that allows this to be done but it must be remembered that not only must the tailstock be moved across by the exact amount but its position must also correspond to the exact length that is required in order to get the precise angle. If we take centres that

have been set over to exactly the right amount and alter the distance between them the angle also changes. Not only then, does the length of the setting need to be precise but so also the depth that the centre protrudes into the hole in the bar. This means that when putting the centre hole into the bar the depth has to be quite precise in order to get an accurate taper.

Getting the tailstock back in line when one has finished with making the taper, is not quite as difficult as it may sound. It can be done by scribing a line across a length of steel and lining both centres against it, alternatively a ruler can be used instead of

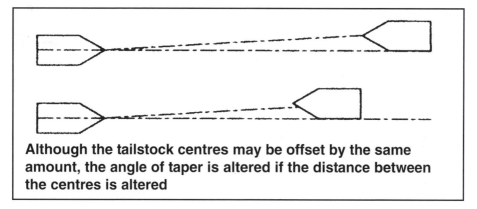

Although the tailstock centres may be offset by the same amount, the angle of taper is altered if the distance between the centres is altered

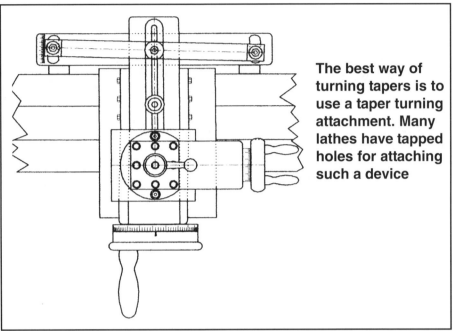

The best way of turning tapers is to use a taper turning attachment. Many lathes have tapped holes for attaching such a device

a piece of metal as long as it has even graduations on both sides. Another way is to drill a hole in a short length of metal and simply line the centres up by pushing them both in to the hole.

To avoid the necessity of setting the tailstock a bar can be made up that has a morse taper shank on one side and a small pointed piece of steel to act as a centre set into it further along on the opposite side.

A useful device that can be home made, although the one shown is a commercial item, is a false centre that is used instead of setting over the tailstock. The use of such a device depends on ensuring that the stock material is of the exact required length any discrepancy will result in the angle of the taper being wrong.

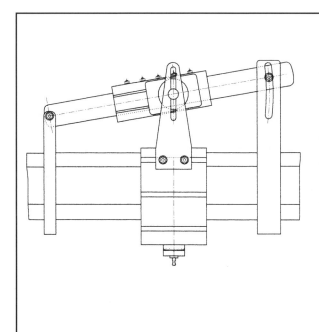

This alternative clamps to the lathe bed. The clamping arms must be pinned and locked to keep them square. Tapers can be machined in either direction

Alternatively the morse taper shank can be dispensed with and replaced with a short length of round bar that can be gripped in the tailstock chuck. We now have what virtually amounts to a set over tailstock and as long as the false centre is lined up to a position exactly on the centre height of the lathe the distance required between the lathe centre and the false one can be calculated.

To machine especially long tapers a special attachment that bolts on to the lathe bed is used. Some lathes have tapped holes especially for fitting such a device, in other instances it can be bolted to the bed. Basically, although there are numerous variations the attachment consists of a long bar that is parallel to the bed and fitted to it is another bar that can be moved to an

angular position the degree of the angle being indicated by graduations; attached to that in turn is a slide. Connected to the slide with a pivot is a slotted bar that is attached to the cross slide of the lathe. The piece to be worked on can either be set between centres or held in the lathe chuck, in the normal way that it would be if it were to be machined parallel. As the carriage is wound forward the tool post which is bolted either directly or indirectly via the top slide, moves at the angle indicated by the graduations, thus giving the required taper. The arrangement has the advantage that there is no need to disturb the tailstock at all.

Taper turning attachments are comparatively easy things to make and over the years several designs of different types have been published in magazines. Perhaps the interest in making them stems from the fact that they require no great skill and are fairly forgiving when it comes to tolerances. The hardest part is to make ones mind up as to how to fit them to the machine. There is a particular difficulty when it comes to fitting them to a lathe with a vee bed that does not have tapped holes for fitting. One either has to make suitable clamps with vees to fit over those on the bed or to make clamps that will actually clear the vees. The latter idea can be achieved relatively simply by fitting the attachment flush against the bed and securing it with something based on a 'G' clamp the body of which goes over the vee. If this idea is used it is essential that good clamps are provided to ensure that the vibration of the machine does not cause them to work loose when in use. Yet anther idea that has been used is to fit the device with magnets, these need to be quite powerful in order to avoid any movement of the attachment whilst in use.

The machining of internal tapers follows the same basic pattern as described for external ones, although it is not possible to use the tailstock set over idea as it is impossible to machine an internal taper if the work is supported with a centre in the tailstock so it will have to be supported by a steady, unless particularly thick bar is being machined, in which case it might be possible to hold the metal in the chuck and machine the taper without the metal flexing at all. This means that only the methods of setting over the top slide, or using a taper turning attachment can be considered.

With the top slide set over it will not be practical to use a travelling steady and the normal fixed steady will severely limit the amount of travel available. This means that some form of improvisation might be called for. One way of doing this is to use a length of wood clamped to the bed in such a way that the edge of it is flush with the work. It is a case really of any thing that will stop the metal moving away from the tool while cutting takes place.

The hole has to be started as for a normal parallel hole and can be drilled with progressively smaller drills to facilitate the actual boring operation. It is possible to purchase tapered drills that will aid the cutting of the taper, but while the expense of buying one of these may be justifiable if a number of identical sockets are required it is probably not worth while for a single job.

Reamers can be bought for finishing morse tapers and it is a good idea to make a reamer for any internal taper that is made. A reamer is only really a tapered 'D'bit and therefore quite easy to make. It does involve turning an outside taper on a length of silver steel but it is easier to turn an external taper than an internal one and a reamer could make a considerable difference to the finish of the taper.

Chapter 14

Measurement and control systems

More accurate and easier to use digital ones are rapidly augmenting tradition forms of measuring equipment. It is so much easier to read the measurement than to have to work it out as one does with a micrometer or vernier gauge. Even if digital equipment has become much cheaper it still costs a considerable amount to obtain enough to cover all aspects of measurement that are likely to be encountered and so no doubt the older methods will have their uses for many years to come.

There are really two forms of measurement needed, the measurements that we are going to work to and measurements of the final result, in fact we use the first to get to the second. Sometimes it is possible to decide on the figure we are working to and get straight there, just checking to ensure all is right when it is finished. In many cases, perhaps most, we need to constantly check where

we are in between by constantly taking measurements of what has been done. For the first we have to work to the graduations on the machine as we have already seen there are three or four dials that we can use. In theory if we know the diameter of the metal that we start with, it should be possible to calculate the figure on the appropriate dial, which is the target for the finished dimension. Given a machine of very high quality and an operator with a great deal of confidence there is absolutely no reason why this will not work. Lesser mortals will get somewhere near the required finished measurement and wish to check to see how things are going and will possibly do this several times before being satisfied with the finished result. As experience is gained there becomes less need for these intermittent checks but it is a brave individual who is willing to dispense with any measurements other than the final one.

It may all depend to some extent on the type of work the individual is doing. Someone who is perhaps restoring an old vehicle and is only going to use the lathe to make nuts and bolts and similar items does not need to be quite as accurate as the person building a delicate instrument where it is necessary to work to a thousandth of an inch or a hundredth of a millimetre, or even smaller amounts. Even the nut and bolt person needs some degree of accuracy and must therefore know how to measure accurately enough for their particular purpose.

Length measurement

Mostly a steel ruler will be used to measure length, be it the overall length or the length that is to be machined. A good steel rule will have graduations fine enough for most purposes, if something more accurate is required, vernier and digital callipers have a tang that will allow measurement to very fine limits. That tang can also be used to measure the depth of holes. One disadvantage of its use is the necessity to ensure that the calliper is perfectly in line with the hole. Small diameter holes do not create too much of a problem as the base of the calliper will straddle the hole, if both sides are touching the edge, the calliper is properly lined up and the tang can be run in the hole to the required depth with complete confidence. If the diameter of the hole is larger than the width of the calliper body only one side will be supported on the edge of the hole. It is here that the likelihood of the calliper being at a slight angle is likely to occur, If the measurement required is critical and as a result of being supported on one side only the tang enters the hole at an angle, there will be a slight

discrepancy in the figure shown on the calliper and the depth of the hole. It is not a very large discrepancy and so may not matter too much, but it will be a different story if absolute accuracy is required.

For quick easy measurement of a hole a depth gauge is probably the easiest thing to use. A simple one can be made by drilling a hole in a piece of metal bar and drilling and tapping a hole to meet the original at ninety degrees. By sliding a length of rod through the bar and securing it with the screw we have a depth gauge. There is more reference to and photographs of depth gauges in Chapter 11. Gauges can be bought although it is probably quicker to make a simple one than to go out and buy one. Modern ones that have a dial read out are also available, the larger of these can be rather bulky but there are some smaller versions with thin probes. Depth micrometers are also a useful way of accurately measuring the depth of holes and these too are available with digital read outs.

Measuring outside diameters has traditionally been done either with micrometers, vernier callipers or outside callipers. Using micrometers or vernier callipers means that an accurate direct measurement is possible but there are occasions when their use is not practical because of limited space. In that case ordinary callipers can be used and are particularly effective when direct comparison is to be made between two diameters. There is a good example of this in Chapter 13. They are also useful where a number of items of the same diameter are required. The callipers can be set in relation to the master and used to verify that the parts being made are the same.

Vernier gauges provide an accurate means of measurement, the photograph shpws to versions, the standard traditional type and one fitted with a dial, which makes reading the measurement a little easier.

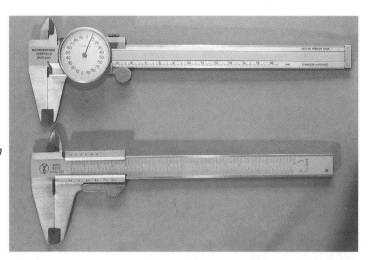

They should be set so that the tiniest amount of friction can be felt when the points are drawn across the surface of the object and it is easy to feel when there is the same tension. Small diameters should be measured when possible with a micrometer and again modern ones have digital readouts that mean it is no longer necessary to calculate the measurement.

Measuring inside diameters, such as in cylinder bores and similar items are a slightly more difficult prospect. Verniers and

A digital calliper frequently referred to as a digital vernier because of its similarity to the vernier gauge which it is the modern equivalent of. It is quicker to read and has instant conversion from imperial to metric measurements. Like the vernier two prongs on the back make it possible to read an internal measurement, although not to any great depth, there is also an extension that slides from the centre to allow it to be used as a depth gauge. This type of instrument can be bought quite cheaply and by removing the arms and making suitable brackets can be used as a digital readout on the lathe.

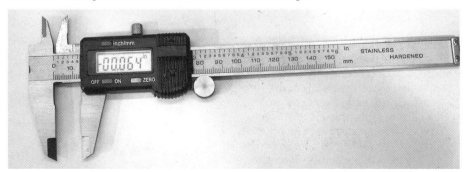

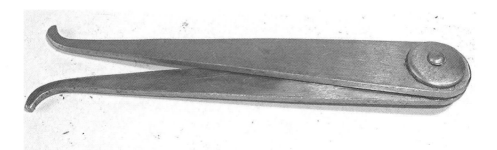

Inside callipers as the name suggests they are intended for measuring inside mobjects. Like the outside version they are frequently convenient in situations where lack of room makes it difficult to use other types of measuring equipment. They are particularly useful for their ability to meause deep inside a cylindrical item.

digital callipers have prongs on the rear face that can be used for the purpose, but they are of necessity quite short and it is therefore impossible to use them to measure a bore of any depth. The old fashioned inside calliper will do the job very well and having been set to the bore diameter, after withdrawal the distance across the tips can be checked with either a vernier, digital calliper or a micrometer. If care is taken it is a very efficient system of measurement. Specially designed digital callipers are made that are designed for inside measurement but their purchase can probably only be justified if a great deal of such measurement is needed.

There are two other items that are

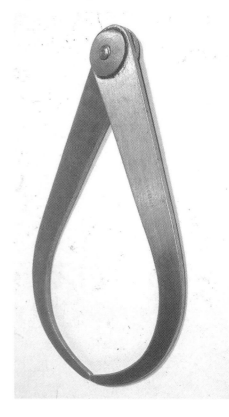

Outside callipers, an ancient measuring device that still has its uses. The size and shape make it convenient at times when vernier and digital callipers are awkward to use. By adjusting the calliper so that just a very slight amount of resistance can be felt when passed over an object it is perfect for comparison of measurements with other items.

Odd leg callipers, more often known as jennies are used for scribing marks parallel to an edge, by using them to scribe four marks at ninety degrees to each other the centre of a bar of metal can easily be found.

probably worth a mention, they are dividers and odd leg callipers, the latter frequently being referred to as Jennies, Dividers are used for measuring between points and for marking off and also for scribing circles. Odd leg callipers are also used for measurement; the plain side is dragged along an edge allowing the point to mark a line parallel to that edge. They are particularly useful for establishing the centre at the ends of large diameter bars.

All the above equipment refers to measuring and checking finished work or work in progress, using modern equipment can mean that this sort of measuring is less likely to be required as digital equipment installed on a machine means that it is possible to work to the very finest of limits. Most people are familiar with digital callipers and if the same principle is used to replace the dials normally found on a machine, the accuracy of the equipment is transferred to that machine.

By far the cheapest and in many ways easiest way to convert a machine so that it has digital readouts on one or all axis, is to fabricate a means of using one of those callipers. Most machines have places where suitable brackets can be fixed and all that is needed to do so is a little ingenuity. Because of the wide variety of machines it is not practical to attempt to offer advice on the best place or how to fix them. There is one problem with using callipers and it is that the jaws can get in the way, the answer to that being to remove them. The steel from which the calliper is made is very hard and an ordinary hacksaw is not going to be able to cope with the job. The best thing to use is one of the very small diamond cutting discs that are sold for use with the small hand held grinders. The disks will do the job quite easily and they can also be used to tidy up any ragged edges that are left.

Using callipers may be the cheapest way to go digital but it is not the easiest, or tidiest and many manufacturers now make a range of digital readouts of varying lengths designed to fit on machines. Most have a bracket of some sort or another to assist in mounting the readout; the type of bracket varies according to the manufacturer. Some supply just a single bracket others supply two. At least one

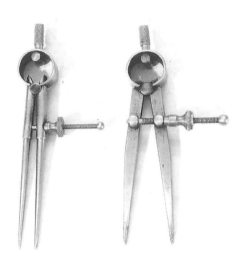

Left: Dividers are well known to most people, the type on the left are known as toolmaker' dividers and have very fine points. The main use is not as commonly believed, for scribing circles but for accurate measurement. Having opened them so that the points locate in the two positions it is required to measure between, the distance between the points can then be accurately measured.

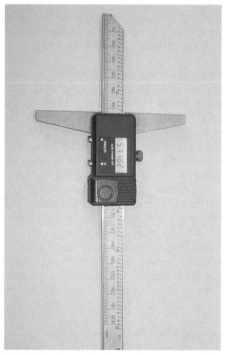

Right: A digital depth gauge, works on more or less the same principle as the digital calliper, this is a 300mm one and while it is ideal for large work probably not the best tool for most work that is done on the home lathe.

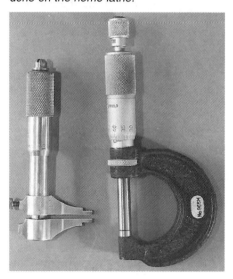

Left: Two forms of micrometer, the one on the left is for inside measurement, a task that is copes with very accurately. The other is a standard type, although the one inch or 25mm type is the more common, micrometers are available in a wide range that can be used for measuring very large items.

A micrometer depth gauge, capable of measuring the depth of a hole with great accuracy, also available are similar tools that have a digital readout instead of the traditional micrometer one.

company does not supply any brackets but the readouts have holes drilled in them ready for screws to be passed through.

If there is a problem at all with the use of these readouts it is that on some the figures are very small and can be difficult to see. If that is the case remote readouts with much larger figuring are available. They are designed to plug into the original readout, manufacturers unfortunately have not made them universal and so in order to do this it is necessary to ensure that both the remote and the original readout come from the same manufacturer. This rules out them being fitted to butchered callipers, but then it is not possible to have it all ways. It is either a case of saving money or getting an improved result.

Digital readouts are available in three types, the normal mechanical/electrical gadgets we also see sold in the form of callipers, a type that works via a wire and another that has a glass scale. The latter two are the most reliable but once again it is a case of getting what one pays for. All three are a great improvement on the graduations on the lathe dials. An additional advantage, whether using digital equipment just for measuring or as adaptations to the lathe is that they can be re-set to zero and also will instantly convert from metric to imperial measurement and visa versa.

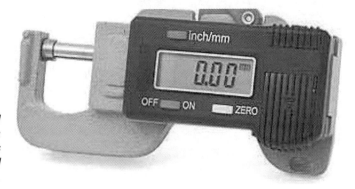

Not the traditional shape for a micrometer but this is the digital version.

159

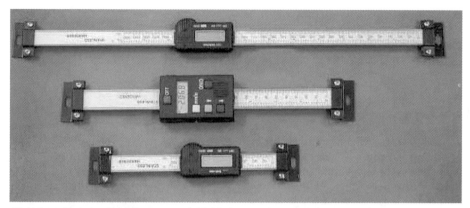

Digital readouts such as these are available in a variety of lengths, actual readout is also available in line or at ninety degrees so that they are suitable for saddle or cross slide measurements. Some come with brackets that can be used for fitting them to the machine.

Some suppliers of digital equipment sell kits that are designed to fit on a particular type of lathe and some modern machines are fitted with digital read out systems as standard. These usually cover saddle, top slide and cross slide, some have remote monitor displaying all three measurements.

Computer controlled machines

There are quite a few computer-controlled lathes on the market; they vary considerably in specification. In industry where production is the object, fully computer controlled machines are very much in evidence. They are not the sort of machines likely to be found in a home

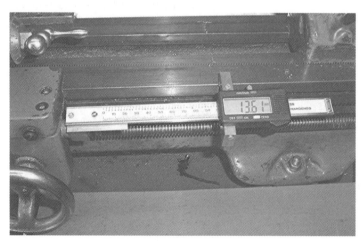

A digital readout fitted to a Myford lathe, two simple brackets were all that was required to enable accurate measurement of the saddle travel.

workshop though. Once set up to do a production run, the doors of the cabinet in which they are enclosed are shut and the machine completes a range of operations that have been programmed in. It is common for one operator to be working, or perhaps watching would be a better word, three or four machines. Certainly not the sort of operation wanted in the home workshop. Might just as well sit indoors and read a book.

As far as lathes that might be considered suitable for the home workshop are concerned there are a number of levels available. The ultimate is one where a Computer Aided Design (CAD) is drawn on a computer and a DXF file, which is an international standard for CAD is placed in the computer this is converted to a format that the machine can read and programmes the machine to carry out all the operations that are on the CAD file. A number of tools have to be mounted on the machine to do the work and these have to be correctly set in order to achieve the required accuracy. Once these are correctly placed the

A single axis readout remote readout box, such boxes are also available for two or three axis measurement. They are attached to the standard digital readout by a cable. The plug on the cable fits a socket on the readout and generally it is essential to get both from the same manufacturer otherwise they may not be compatible.

A two-axis readout system with remote monitor fitted to a Myford Lathe gives accurate measurement to saddle and cross slide, with an easy to read monitor in full view of the operator. Photograph courtesy of Allendale Electronics Ltd.

operator can sit back and allow the machine to make the component As long as the initial work has been carried out correctly the work will be absolutely accurate and the part can be repeated over and over again.

The idea is very nice and if the object is to make something as accurately as possible, as quickly as possible and there is plenty of money to spend it is ideal. Some may feel however that the enjoyment of

machining is in using ones own skills and that is not the way to do it.

If we come down the scale a notch it is possible to have a machine that will carry out a number of operations, not by reference to a CAD disk but by being programmed by the operator. The machine works in the same way as the more advanced machine but the operator enters the measurements required on a keypad. The tools travel the distances and depth he or she has put in, thus the accuracy is still there but things are now more under personal control. The operator has had to set the tools up correctly, set up the metal correctly; the only thing taken out of his or her hands has been the actual machining operations. There is still plenty to do to get things right, but it is far easier and more accurate are than the old way.

A modern lathe by WD Ternick features a digital readout system as well as screw cutting gear box and geared head as standard equipment.

A Hardinge Lathe, partially CNC controlled, with steppe motors to the control handles but also capable of hand operation if required.

We can even come down a shade lower in the scale yet and obtain a machine that will carry out one operation at a time. To machine along the length of a bar the operator sets up the tools and the metal, winds the tool in to the depth of cut required. Puts in the keyboard the distance along the bar to travel and starts it going. The cross slide can also be programmed for length of cut, the result is a machine that gives a great deal of accuracy, while allowing for a limited amount of skill.

Let us then have a look at how computer control works and its possible uses. The spindle is driven in the normal way by a standard type of electric motor, the saddle and cross slide and in some instances the tailstock are driven by stepper motors. A stepper motor does just what the name suggests rotates in a series of minute steps, there are three types of these and should anyone be considering making their own machine the hybrid motor is the one to use. The motors have to be driven by special controllers and they impart travel to the required amount. To avoid backlash and give a smooth operation the slides are driven by ball screws instead of the more usual type. Ball screws are high precision components with specially ground threads and nuts that can loosely be described as like a ball race. They give a very smooth operation and a great deal of torque.

A console for a fully CNC controlled lathe, operation is by entering the required movement from the control panel.

Quite a number of home machinists have converted lathes to some form of computer control, this means fitting the stepper motors linked to their necessary control units, to the main lead screw and that of the cross slide. There are some

An Emco Compact 5 lathe that is fully CNC operated.

instances where the lead screws have been changed and ball screws fitted, others have gone ahead using the normal lead screws. Results so far appear to have been very satisfactory, although the set ups are sometimes a little bulky. Perhaps this is a small price to pay for the added interest gained by doing the conversion as well as that gained when using the machine.

It can be seen that modern methods are now changing lathe work from they way it was once known and although there will always be die-hards who like the challenge of sorting it all out, it is inevitable that some form of digital control will be the way ahead in the future.

Should the reader be interested in pursuing computer controlled machining the S.I. Model Books publication "Electromechanical Building Blocks" by Pat Addy ISBN 9781854862433 provides a wealth of information and practical electronic circuits.

Chapter 15

Miscellaneous operations - 1

The various operations that have already been dealt with could be described as the basics, there is a great deal more that can be done on a lathe and it is well within living memory that folk used the machine for every possible operation. No doubt there are still some that do so but most people these days have a little more equipment and describing every possibility would take a small library not just a single book, so we will deal with just a few of the extras.

Fly Cutting

The term fly cutting relates to a process used for machining large diameters, although it is more commonly used on a milling machine it is a valuable process on the lathe as well. The type of fly cutter used

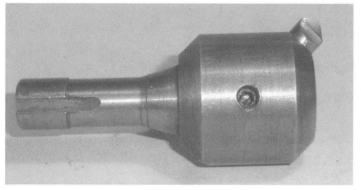

A flycutter of the type held in the lathe chuck, This type of cutter is effective but because of its limited swing can only cover small areas.

Fly cutters that are used on a faceplate should always be counterbalanced and seen here is a cutter together with its counterweight.

on the milling machine can also be used on a lathe and held in the chuck. In addition a type that is an attachment to the faceplate can also be used and often will prove to be more useful as they swing in a wider arc. This suits the method, which relies on a tool covering the whole of the work in a single sweep, in order to get a good finish.

The rear of this flycutter has a raised section that is an exact fit in the slots of the faceplate, this stops the cutter from moving round as it strikes the work.

Basically a fly cutter is an ordinary cutting tool as used in the lathe tool post, set in a special holder. As well as using specially designed tools for fly cutting it is possible to put an ordinary turning tool off centre in the four-jaw chuck and use that.

Work to be fly cut will normally be clamped to either the saddle or cross slide, preferably the former, it may be necessary to use the top slide for angular work. This should only be done as a last resort. Usually fly cutting involves intermittent cuts that have a heavy blow and there will be less

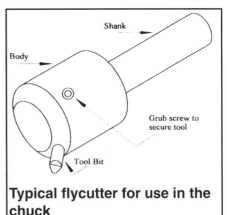

Shank

Body

Grub screw to secure tool

Tool Bit

Typical flycutter for use in the chuck

Before flycutting a flat surface it is essential to ensure that it is square to the faceplate and this can be done by placing a block between thework and the faceplate and adjusting it before clamping.

Having ensured the casting is square flycutting has now commenced.

A large casting clamped to the cross slide for flycutting, note in this case no matching counterweight was available and so a large heavy washer was stuck to the faceplate to do the job.

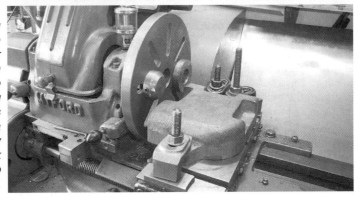

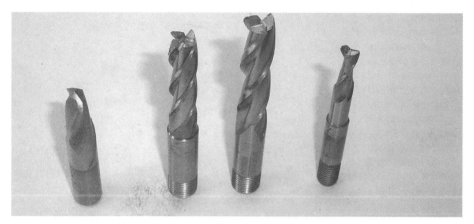

Cutters that are used for milling. On the left is a ball nosed cutter used for milling a small radius, next to it a four flute cutter, then a three flute type, with a two flute slot drill on the right. Although called a slot drill, the term is not to be taken literally as the cutter is not intended for drilling, but for cutting slots which it does more efficiently that the other types of cutter. The three flute type is a compromise between a slot cutter and end mill. The end mill with four flutes is designed for making flat surfaces.

strain on the machine if the saddle is used. It hardly needs to be pointed out that work must be absolutely secure before cutting operations commence. It is also necessary to have in mind that the blow when the tool strikes the work will create a strain on the main bearings and so cuts should be kept as shallow as possible. A great deal also depends on the position of the cutting tool. If it is rotating at the extreme edge of the faceplate a greater strain will be imposed on the machine than that given by a tool only rotating at an 1in or 25mm diameter. Because the action and position of the tool is eccentric it is advisable to counterbalance the faceplate by mounting a large nut or a piece metal on the faceplate, opposite the cutter.

The process is suitable for any work where it is necessary to generate a large flat surface and it is possible if care is taken to get a superb finish. The larger the diameter of the cutter movement the slower the speed and feed should be, at any time, fly cutting is a slow operation and must not be hurried. As with all machining operations it is essential that the tools used are sharp.

Milling

These days a large majority of home workshops have a milling machine as part of the equipment, there are still a few occasions where it might be found more convenient to use the lathe for a little bit of milling. The cutters that are used are normal three or four fluted milling cutters and two fluted slot drills.

Mostly a vertical slide is used to manipulate the work, as implied by the name the slide has a vertical movement. Add this to the horizontal movement of the cross slide and we have the same set up

A vertical slide which is the usual device used for milling operations. It is mounted on the cross slide this gives two axis, across and up, depth of cut can be adjusted by moving the cross slide in or out. After adjustments have been made it is essential that all movements that are not going to be used for that particular operation a locked tight to prevent them moving. A lot of pressure is involved in milling operations.

as a vertical milling machine, except that the set up is at ninety degrees to what one would get from the miller. Vertical slides are available in two types, plain and swivelling, the latter being pivoted in the centre so that the table can be set to an angle if required. They also swivel from the base, but that

can be done with the plain slide anyway. This would of course be for angular work and if that is to be done the angle of the slide should be checked with a protractor. Certain small tasks can be carried out without the use of a vertical slide, by simply bolting the work to an angle plate. There is no vertical movement available in that case, but for a task such as making a keyway or putting a small flat on a round bar that is of no consequence. If it becomes necessary

In order to give added support to the vertical slide packing has been placed underneath. Note the use of aluminium strips to prevent damage to the lathe bed.

169

The change wheels of the lathe can be used for simple division, a detent can be filed from mild steel and by using a simple clamping arrangement, how to make it will depend on the lathe, so it will fit between the teeth of the gears, the lathe can be rotated and any number of division obtained as long as the wheel in use is exactly divisible by that number.

what the machine is designed for. It is necessary to take fine cuts as even the most robust of machines will suffer from vibration if the work is too heavy. One good idea is to support the bottom of the vertical slide, or the vice if it is easier to do so and this can be done with small jacks, or alternatively by packing with blocks.

Dividing

Dividing is more common on a milling machine than on the lathe, nevertheless there are plenty of occasions when accurate division is required. It might be in order to make some form of cutter, a series of screw holes for securing a part to another, or perhaps filing flats in order to make a key. The equipment necessary for dividing largely depends on the work that is undertaken. A typical example of simplicity in order to obtain three equal divisions is to put a metal block under one jaw of the three jaw chuck and mark the work off, then rotate the chuck by one jaw and mark off the second position, then do the same for the third. The same principle could be applied to work held in a four-jaw chuck.

Another popular method is to use the change wheels of the lathe, make a split mandrel to allow a wheel with a suitable number of teeth to be mounted securely in the lathe spindle and make a detent

to do milling on the lathe and no vertical slide is available, the top slide can be mounted vertically on the cross slide by bolting it to an angle bracket, or perhaps a pair of them, depending on what is available. The only real purpose of the vertical slide is to give vertical movement, which the top slide upended can do just as well. All other movement of the work will be via the lathe carriage and cross slide.

Work can be bolted directly to the vertical slide, using tee bolts. Most slides have slots that are suitable for this, but there are examples that do not, these are usually designed for the smaller lathes and have a series of tapped holes set in strategic places. It is also possible to mount small machine vices and use them to hold the work, Whichever way it is done it is essential that the work is secure as milling operations create a lot of sideways thrust which must be born in mind when setting up the work.

A lot of good quality milling has been carried out using the lathe even if it is not arrangement that will fit somewhere firmly on the rear of the lathe. The usual place for this is the quadrant that is used for fitting the change wheels. If say a forty tooth wheel has been chosen it is possible to get divisions of any number that will divide by forty. Suppose one wants eight divisions, move the wheel five places for each division, four divisions then move it round ten teeth for each one and so on. By using different change wheels it is possible to get a very large number of divisions and all obtained in a very simple fashion.

These simple methods have their advantages but are not always suitable for operations like clock wheel cutting when the number of divisions required are not equal to any change wheel that one has, in that case a dividing head is needed. A number of designs have been produced for the home manufacture of dividing heads and they are also available as commercial items. In the case of the latter they are sold as universal tools, designed to fit on a lathe or milling machine. As this is a book about lathes we will stick with a description of a dividing head as used on the machine. The basic requirement is a worm and wheel, the ratio of which ideally should be forty or sixty to one. Assuming the lathe has change wheels, all that is required is a worm to mesh with these. It needs to be on a shaft and there must be some method of securing the shaft to the lathe in such a way that it can be rotated. Most people use the frame used to hold the change wheels for this purpose, but dividing heads have been produced where the tapped hole used to hold a sight feed oiler has been used. In the first case the set up will be horizontal and the second scenario would result in a vertical one.

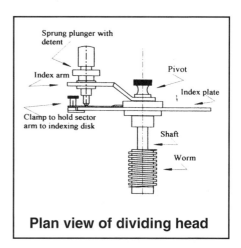

Plan view of dividing head

The change wheel, which we will assume in this case has forty teeth, must be secured to the lathe mandrel and the worm adjusted to fit. It then follows that one complete rotation of the shaft attached to the worm will result in the mandrel rotating one fortieth of a revolution. So we already have a means of dividing, by simply counting the number of revolutions of the worm and if eight divisions are required the worm has to be turned five complete revolutions for each division. We can of course also get that by using a simple detent with the same wheel. What if the division required is not nice and neat and will not divide exactly into the number of teeth on a change wheel? There is then a need for more flexibility, to get this an index plate has to be fitted onto the worm shaft. The index plate is a circular disk with a number of holes drilled in it and because there is plenty of space on the disk we have not one circular row of holes but four or five.

The disc to be used is fitted to the bearing, not the actual shaft; this means the shaft can be rotated while the index

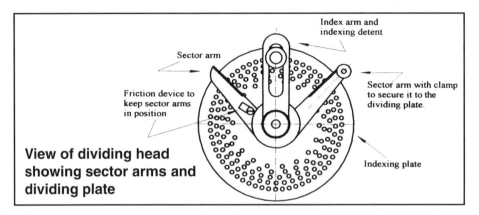

Sector arm

Index arm and
indexing detent

Friction device to
keep sector arms
in position

Sector arm with clamp
to secure it to the
dividing plate.

**View of dividing head
showing sector arms and
dividing plate**

Indexing plate

plate remains stationary. Fixed to the shaft is a narrow plate holding an index pin that fits into any selected hole in the plate. The shaft is slotted so that the pin can be moved up and down in order to get it aligned with any of the rows of holes. The pin is spring-loaded and the knob used to release it also acts as the handle to rotate the worm. Assuming that we need to divide our item into six different sections, we find that six will not divide exactly into forty, the required number is $6\text{-}\frac{2}{3}$ note we work in fractions not decimals. To get this a plate with a number of holes, divisible by three is needed and for the sake of convenience we will take thirty. Twenty spaces along the row of holes will be two thirds of a revolution, which is exactly what is required. To get the number the worm must be rotated completely six times, plus the twenty divisions indicated by the holes in the plate.

To keep counting twenty divisions round the plate for each division is not only time consuming but extremely irksome and

with odd numbers it seems to be even worse. Included in the set up therefore are two plates known as sector plates. They are constructed in such a way that although mounted one on the other, the actual arms are at the same level. Sometimes one is fitted with a clamp that allows it to be secured to the sector plate. In addition a friction device is interposed so that they will remain in the position to which they have been set. The friction device varies from manufacturer to manufacturer, some fit a screw arrangement, others a spring clip. The arms of the sector plate are set to cover the required number of holes, in this case twenty, these are clear spaces, and the hole in which the detent is fitted is not counted. To obtain the odd fraction of turn, the pin is simply rotated from one sector arm to the next after completing the required number of full rotations. The arms can then be moved, using the same setting, so that the first one touches the detent pin and the operation repeated and so on.

Chapter 16

Miscellaneous operations – 2

Hand turning

The expression hand turning is generally applied to the use of a cutting tool that is not fixed in a tool post; instead a special rest is used. There is nothing very complicated about a hand turning rest, it comprises of a base that is bolted to the cross slide, a piece of flat bar forms the top section and is fitted to the base via a pillar that can be adjusted for height. Usually but

not always the top bar will be angled downwards slightly where it faces the work this allows some clearance for the swarf to get away. Tools have a long shaft, usually somewhere between six and nine inches (150 and 225mm). However the length is not mandatory and some operators like tools that are considerable longer than that. Firmly fitted to the opposite end of the tool is a handle, usually but not always this is made from wood.

Operation is simply a matter of applying pressure with the tool, while the metal is rotating. It is necessary to keep the rest as close to the work as possible in order to avoid the chance of the tool snagging between work and rest, something that can

A hand tool rest, these are usually associated with wood turning operations but have their uses for the metal worker and they are extensively used by clockmakers who do a lot of hand turning.

Hand turning a small pillar for a model steam engine. Note the well gloved hand for protection in case of an accident.

always a good idea to wear a stout pair of leather gloves when carrying out this type of operation.

Forming radii

The use of form tools for the manufacture of small diameter radii and other small shapes has already been dealt with. We cannot though use a form tool if a large radius is required and so other methods are called for

There is simple tool that can be used for making reasonably large radii. It is like a short piece of tubing, made from high carbon steel and hardened and tempered. Silver steel is the best material to use and it can be obtained from any good engineering tool stockist. As well as the working part, which can be fairly short in length an extension and handle are needed. It is not practical to give an overall length for the tool as like the hand turning tool, which it really is, it will depend on individual taste. What can be said is that it is necessary to apply leverage when working with it, so it must have sufficient

obviously be quite dangerous if it happens. The operation is exactly the same as that done by the wood turner, but removal of material is more difficult when dealing with metal. When turning by hand an important thing to remember is the tool angles, it is only too easy to move the tool to a position that will give it a negative rake. It is common practice to finish hand turned work, using a fine file and abrasive cloth or paper. It is

A tool for making a radius, it is used in conjunction with a hand rest and is a simple way of making curved surfaces.

174

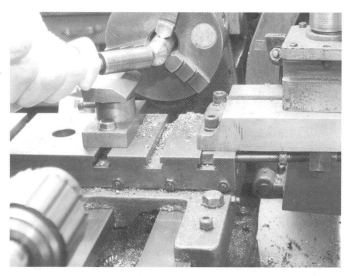

The radiusing tool in action, note that the turning rest has been set so that the edge is just about centre height.

length to give that. It is used by applying it to the rotating work and moving the tool round at the same time. Good support for the tool is needed and it is best if a hand turning rest is used. If one is not available, the edge of a four-way tool post might do the job, but only after all the retaining bolts have been moved.

Inevitably when hand turning and using this tool, there is an element of danger that the tool will catch in the work and snatch. It is therefore essential that only light pressure is used and no effort made to remove a large amount of metal in one single movement,

The final tool used for radiusing is not surprisingly, called a radius tool and consists of an ordinary turning tool set in a holder that can be swung through an arc. A number of designs for such tools have been published, some lighter ones work from the tool post of the lathe. More heavy-duty

Photograph showing the finished job.

versions are mounted directly on the cross slide, both types are usually lever operated, although there are examples that are designed to be wound. These are more common where very large radii are to be machined. The diameter of the radius depends on the distance the pivot is from the work, so there has to be some means of adjusting that distance.

A selection of knurling wheels that range through a number of different patterns.

Apart from when a form tool is used, it is always wise to start the radius off by removing the edges to be radiused by gently machining them off with a combination of shallow cuts using the cross and top slide. Trying to work directly without this initial preparation is extremely difficult and can be dangerous whenever one is using a lever operated device.

Knurling

Knurling is the name given to putting impregnated patterns on work, it serves two purposes. Firstly it can considerably improve the appearance and secondly it provides a good grip on the surface and is mainly used on tools and similar things. It is a very simple operation requiring nothing more than the application of specially hardened wheels that have been impregnated with the required pattern to be pushed or squeezed into the work. The tools for doing so are easy to obtain and are equally easy to construct. Well, the

Two examples of home made knurling tools. Both of the calliper type and both equally effective.

A knurling tool designed to work from the tailstock. As well as not imposing a strain on the lathe bearings it also leaves the tool post available for other purposes.

body that holds the wheel is, the actual wheels are a very different proposition and are best purchased. There is a wide variety of types of holders and the choice of which type to use must be a personal one. The simplest form is nothing more than a bar of metal with a slot and cross-drilled to accept a pin. The wheel is placed in the slot and a pin pushed through to act as an axle. It is a very easy tool to make but there is one thing to watch for when doing so. As has already been mentioned the wheels are extremely hard and if mild steel or something similar is used for the pin, it will literally only take a few minutes for the wheel to wear through the pin and the whole lot to come to pieces. There are two ways to avoid such a disaster, the first is to make the pin that is used hard as well and the second is to make it of phosphor bronze. This is sticking to the standard practice for bearings where one component will be hard and the other soft, or both extremely hard.

Instead of a single wheel the use of two is more common, particularly where a diamond pattern is required, each wheel has one half of the pattern and between them they do the finished job. They are held in a floating head that will allow both wheels to settle nicely on the work, where as a fixed head unless very carefully set up is likely to apply more pressure on one wheel than the other. Both these types of tool have their

uses but whether or not they are suitable for use in the home workshop is another matter. In order to do their work they must be pushed very hard into the work, which has the effect of pushing against the bearings of the lathe. On a big industrial machine this may well be of no consequence, with the much lighter machines that are usually found in the home workshop it might possibly create undue wear. If one of these types is to be used, the usual precautions must apply by working as close to the chuck as possible,

A home made filing rest, the use of one of these ensures a flat surface and is idea for work such as putting square ends on shafts.

thus avoiding the leverage that one gets as we move away for the support.

A far better proposition for the home workshop is to use what is generally called a calliper-knurling tool and again they are readily available for purchase if one so wishes. They are not at all difficult to make and can possibly be made in the time it takes to get to the nearest tool shop to buy one. Because they clamp the wheels on either side of the work, they put no strain whatever on the bearings of the machine.

Knurling wheels are available in various diameters and variation in pattern. Perhaps that last remark should be clarified a little as generally speaking we are only talking of two patterns, straight or diamond. The variation comes in the spacing and depth of that pattern and they can be fine, medium or coarse. In industry it is usual to have wheels that are designed not only for different purposes but also for different metals and diameters thereof, the home worker is hardly going to need such refinements and most people will be satisfied with one or two pairs.

Knurling can hardly be referred to as an art it is a case of deciding where the knurl is required, clamping the wheels round and rotating the lathe. The speed of the lathe should be very slow; in fact back gear is advisable. Don't try and tighten the knurls too much and move the tool along very slowly. Moving too fast will destroy the pattern. Assuming the depth required has not been achieved on the first go, do not remove the wheels from the work, but keeping the lathe at the same slow speed move the knurls in the other direction with a little more pressure applied. The process can be repeated as often as one wishes until the desired result is achieved. Do not

decide to put the lathe in reverse because the knurls are being wound in the opposite direction. Knurling can create a great deal of heat, particularly if a long stretch of metal is being knurled, so apply a good quantity of cooling fluid as long as the work is being done.

At exhibitions one is likely to overhear people discussing the finer points of knurling on exhibits, most of what is said is a load of rubbish. There are only two things to watch for, assuming the pattern is correct, that is to try and avoid wide flats on the top of the knurling pattern, this is caused by the wheels not going deep enough. The other thing to be avoided is the exact opposite, don't put the wheels in so deep that the top of the pattern feels sharp, the pattern is there to do a job, which is to make the artefact easy to handle. If the edges of the pattern are like knives cutting into ones hands that object has not been achieved.

One final point: if for some reason the wheels have been lifted from the work and it is then decided that the pattern is not deep enough, all is not lost. Close them very lightly on the original pattern while rotating the lathe by hand, the gentle movement achieved by rotating it by hand will allow the wheels to relocate in the original indentations. It requires care but at the same time is not difficult, putting the machine under power and just squeezing the wheels in will most probably result in a double pattern and the job will be spoiled.

Filing

It is frequently the case that a small flat section is required on a piece of work and that it is not convenient or maybe even not possible to mill it. One good reason is that

Known as a Jacot Tool or Jacot Drum this is a useful filing aid. A number of grooves are available to support different diameters. It is a tool that is particularly of use to clock makers.

it may not be desirable to remove the piece from the lathe. Of course one can simply move the tool post out of the way and file away to ones heart content, the end result is unlikely to be smooth and even but perhaps that will not matter. To get it right it is necessary to use a guide of some sort; the most sophisticated one being a filing rest. This consists of two rollers on a frame that is adjustable for height. It might be possible to buy one but they are very hard to find. One reason being that they need really to be made to fit a particular make of lathe. The height and spacing for bolting them down will differ from manufacturer to manufacturer so they are not a very practical proposition commercially. They are not difficult to make and are an interesting little project for a rainy afternoon.

Filing flats on small diameter material is always a problem as invariably the metal starts to bend away from the pressure of the file. The answer to this is to drill a hole in a short length of bar, of the same diameter as the piece that needs to be

worked on and cut part of it away. We are now left with a solid support on which to work at least for the first two filing operations, depending on how much material is to be removed. If the bar is no longer going to give support drill a smaller hole in the other end and repeat the operation so that it does. This is an idea that clockmakers have used for years but they make a Jacot Tool, or Jacot Drum that has a series of grooves that can be used. Finally one little dodge that works quite well is to support the work with blocks, packed up from the lathe bed, it is best in this case if wood is used so that the work does not get marked.

Keyway Cutting

Keys are used as a means of both locating and retaining components, there are mainly two types of key and keyway that we are likely to come across, the flat type and the Woodruff Key. The latter is not found all that often but is basically a piece of flat steel that is radiused and it fits into a radiused

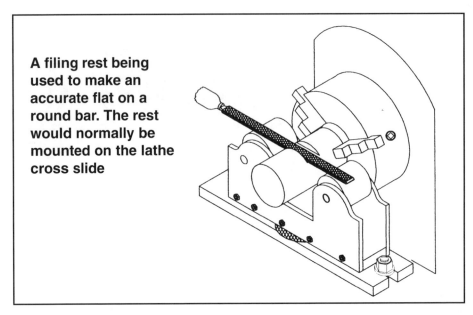

A filing rest being used to make an accurate flat on a round bar. The rest would normally be mounted on the lathe cross slide

slot in a shaft. It is used in instances where a normal key is not practical, having the advantage that the key can be located between the shaft and component after assembly and does not require the keyway to be cut right to the end of the shaft. It is an effective device but sometimes it can be difficult to set the key in its location, this can result in it falling and as the keys are very small they can be difficult to find. It is therefore best to keep a couple on hand in case of emergencies.

Keyways for woodruff keys can be made by simply plunging a rotating milling cutter (not an end mill or slot drill) or slitting saw of the correct diameter and thickness into the shaft for the required depth. A woodruff key can be made by slicing a piece off a round bar of suitable diameter and then cutting across the slice with a hacksaw. The keyway for a woodruff key is only in a shaft, not in the component to be mounted thereon that has an ordinary standard keyway.

The regular type of keyway and key consist of a straight slot in a shaft that matches a similar slot in another component. The key is a short length of steel, usually square in section; the keyway in a shaft can be milled or planed out in the lathe. To plane it a tool of the required shape is put in the tool post and racked along. It is a laborious business but fortunately a task one does not need to do too often.

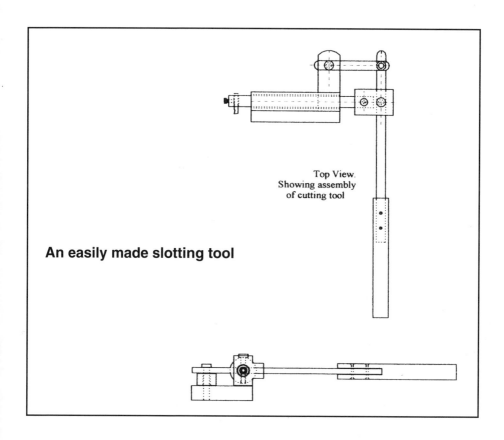

Top View.
Showing assembly
of cutting tool

An easily made slotting tool